Reading Aristotle

PHYSICS VII.3

Reading Aristotle

PHYSICS VII.3

"What Is Alteration?"

Proceedings
of the European Society for Ancient Philosophy Conference
organized by the HYELE Institute for Comparative Studies

Vitznau, Switzerland, 12–15 April 2007

Edited by

Stefano Maso, Carlo Natali, and Gerhard Seel

PARMENIDES
PUBLISHING

Las Vegas | Zurich | Athens

PARMENIDES PUBLISHING
Las Vegas | Zurich | Athens

This edition published in 2012 by Parmenides Publishing
in the United States of America

ISBN soft cover: 978-1-930972-73-5
ISBN e-Book: 978-1-930972-74-2

Library of Congress Cataloging-in-Publication Data

Reading Aristotle: Physics VII.3 "what is alteration?" : proceedings of the European Society for Ancient Philosophy conference : organized by the HYELE Institute for Comparative Studies, Vitznau, Switzerland, 12/15 April 2007 / edited by Stefano Maso, Carlo Natali and Gerhard Seel.
p. cm.
Includes bibliographical references (p.) and index.
ISBN 978-1-930972-73-5 (pbk. : alk. paper) -- ISBN 1-930972-73-3 (pbk. : alk. paper) -- ISBN 978-1-930972-74-2 (e-book : alk. paper) -- ISBN 1-930972-74-1 (e-book : alk. paper)
1. Aristotle. Physics. Book 7--Congresses. 2. Science, Ancient--Congresses. 3. Physics--Early works to 1800. I. Maso, Stefano. II. Natali, Carlo, 1948- III. Seel, Gerhard, 1940- IV. European Society for Ancient Philosophy.
Q151.R43 2011
530--dc23

2011050131

Typeset in Trump Mediaeval and OdysseaUBSU (Greek)
by 1106 Design | www.1106design.com

1-888-PARMENIDES
www.parmenides.com

We dedicate this volume to the abiding memory of Michael Frede.

The Vitznau Conference was the last of the many "Reading Aristotle" gatherings he attended.

Contents

Acknowledgments

In this volume we present the results of the conference on Aristotle's *Physics*, book VII, chapter 3, which took place in Vitznau (Switzerland) from the 12th to the 15th of April 2007. The conference was organized by the European Society for Ancient Philosophy (ESAP) and the HYELE Institute for Comparative Studies. We wish to thank the latter for its generous sponsorship and the organizational support which made the conference possible. We also want to express our gratitude to Parmenides Publishing who agreed to publish this rather unusual and innovative work. We did not simply gather the papers as presented at the conference, but continued our common work on this difficult text, incorporating and harmonizing the input from all participants in preparing this publication.

The following scholars participated in the Vitznau conference, and we thank them for their valuable contributions to the debate: István Bodnár, Damian Caluori, Tomás Calvo Martínes, Walter Cavini, David Charles, Sylvana Chrysakopoulou, Ursula Coope, John Cooper, Klaus Corcilius, Catherine Darbo-Peschanski, Frans A. J. de Haas, Wolfgang Detel, Theodor Ebert, Michael Frede, François Grandjean, Edward Halper, Arnold Hermann, Fritz-Gregor Herrmann, Katerina Ierodiakonou, Doukas Kapantais, George Karamanolis, Vassilis Karasmanis, Inna Kupreeva, Jean Louis Labarrière, Anna Marmodoro, Stefano Maso, Fabio Morales, Benjamin Morison, Carlo Natali, Catherine Osborne, Yamina Oudai Celso, Michail Peramatzis, Christian Pfeiffer, Vasilis Politis, Oliver Primavesi, Spyros Rangos, Christof Rapp, Dory Scaltsas, Gerhard Seel, Miira Tuominen, and Cristina Viano.

Stefano Maso, Carlo Natali, Gerhard Seel

List of Contributors

István Bodnár is Associate Professor of Philosophy at Eötvös University and Central European University, Budapest. He is the author of several articles and co-editor of *Eudemus of Rhodes* (2002).

Ursula Coope is Professor of Ancient Philosophy at Oxford University, and a Fellow of Corpus Christi College. She is the author of *Time For Aristotle: Physics IV.10–14* (2005/2009).

Frans A. J. de Haas is Professor of Ancient and Medieval Philosophy at Leiden University. He is the author of *John Philoponus' New Definition of Prime Matter: Aspects of its Background in Neoplatonism and the Ancient Commentary Tradition* (1997) and co-editor of *Aristotle's On Generation and Corruption I* (2004) and *Interpreting Aristotle's Posterior Analytics in Late Antiquity and Beyond* (2010).

Stefano Maso is Assistant Professor of History of Ancient Philosophy at Università Ca' Foscari, Venice. He is co-editor of *La catena delle cause. Determinismo e antideterminismo nel pensiero antico e in quello contemporaneo* (2005) and the author of *Capire e dissentire. Cicerone e la filosofia di Epicuro* (2008), and *Filosofia a Roma. Dalla riflessione sui principi all'arte della vita* (2012).

Benjamin Morison is Associate Professor of Philosophy at Princeton University. He is the author of *On Location: Aristotle's Concepts of Place* (2002), and a contributor to the Cambridge Companion to Galen.

Carlo Natali is Professor of History of Ancient Philosophy at Università Ca' Foscari, Venice, and Visiting Fellow at All Souls College, Oxford. He is the author of *The Wisdom of Aristotle* (2001), and co-editor of *La catena delle cause. Determinismo e antideterminismo nel pensiero antico e in quello contemporaneo* (2005).

Oliver Primavesi is Professor of Greek Philology at Ludwig-Maximilians-University, Munich. He is the author of *Empedokles Physika I: Eine Rekonstruktion des zentralen Gedankengangs* (2008), and co-author with Alain Martin of *L'Empédocle de Strasbourg.* Introduction, Édition et Commentaire (1999).

Gerhard Seel is Professor Emeritus at the Institute for Philosophy, University of Bern. He is the author of *Sartres Dialetik* (1971), *Die Aristotelische Modaltheorie* (1982), and co-editor of *Ammonius and the Seabattle: Texts, Commentary, and Essays* (2000).

Cristina Viano is Senior Researcher (Directeur de Recherches) at the French National Center for Scientific Research (CNRS), Léon Robin Center, Paris-Sorbonne. She is the author of *La matière des choses. Le livre IV des Météorologiques d'Aristote, et son interprétation par Olympiodore* (2006), and editor of *Aristoteles Chemicus. Il IV libro dei Meteorologica nella tradizione antica e medievale* (2002), and *L'alchimie et ses racines philosophiques. La tradition grecque et la tradition arabe* (2005).

Robert Wardy is Reader in Ancient Philosophy and Fellow and Director of Studies in Classics and in Philosophy at St Catharine's College, Cambridge. He is the author of *The Birth of Rhetoric: Gorgias, Plato and their Successors* (1996), *Aristotle in China: Language, Categories and Translations* (2000), and *Doing Greek Philosophy* (2005).

Foreword

Stefano Maso

This volume contains the first tangible results of group research carried out by a team of Aristotelian scholars, who first came together in 1995, generously hosted by Vassilis Karasmanis (National Technological University, Athens) at the European Cultural Centre of Delphi. The evocative setting undoubtedly increased the participants' enthusiasm and contributed to the initiative's success. Since then, regular meetings have taken place every spring. Among the earliest promoters was David Charles (Oriel College, Oxford); he was joined in 1998 by Carlo Natali (Università Ca' Foscari, Venice). Gerhard Seel (University of Bern) became part of the organizing committee in 2003, and J. L. Labarrière (CNRS, Paris), who had attended the meetings since 1997, joined him in 2005.

The main purpose of the meetings is to read and interpret, line by line, a short but important chapter of Aristotle's works. In this way, attention is focused on key texts of particular exegetical and theoretical interest. The chosen chapter is divided into shorter passages and a whole session is dedicated to each passage. The sessions start with the presentation of a translation and a first analysis of the main problems, which then become the subject of an intense debate that illustrates the different schools of thought and methodological approaches. Over the years, the confrontation of these different points of view has had a beneficial effect on scholarship and stimulated research activity worldwide.

The type of study in question is best illustrated with reference to the texts read and studied over the years (in chronological order): *De anima* III.7 (desire and thought in relation to action); *Physica* III.2 (defining motion by modalities); *Analytica posteriora* II.19 (knowledge of first principles by induction); *Analytica posteriora* II.10 (definition of substance); *Ethica Nicomachea* VII.7 (pleasure and pain in relation to self-control); *Ethica Nicomachea* VI.2 (sensation, thought, and desire in relation to action); *Metaphysica* IX.7 (being, potentially and actually); *Metaphysica* XIII.10 (separable ideas and substances, the difficulties of this position); *Metaphysica* II.2 (the necessarily limited number of causes); *De partibus animalium* I.1 (method of research in natural science); *De anima* III.10 (appetite and thought as sources of movement); *Topica* I.8–9 (categories of predication); *Physica* VII.3 (what is alteration?); *Ethica Nicomachea* VI.13 (moral excellence and practical wisdom); *Metaphysica* I.7 (the principles and causes of reality according to the ancient philosophers); *Politica* I.2 (the origin of the state); and *Physica* II.5 (accidental causes). In 1998, a supplementary session was organized in Venice, where David Charles ("Happiness and Analogy in Aristotle"), Vassilis Karasmanis ("Possibility in Greek Philosophy"), and Gerhard Seel ("Aristotle's Original Insight") presented papers.

In order to sustain the ongoing work by a stable organization, in 2005 the scholars decided to found the European Society for Ancient Philosophy (ESAP). Since then, the annual meetings have taken place in Athens (2006, 2008), Vitznau (2007), Delphi (2009), and again in Athens (2010, 2011). The 2007 meeting, hosted by Gerhard Seel and organized with the support of Arnold Hermann, director of the HYELE Institute for Comparative Studies, took place in Vitznau, Switzerland, on the banks of the beautiful Lake Lucerne. The group decided to publish the results of this meeting in order to make them accessible to a wider audience of scholars and students. The present volume is the fruit of this common effort.

The Vitznau meeting examined one of the crucial texts in Aristotle's theory of change, in which the philosopher deals with the question what alteration is and what it is not. Aristotle discusses change in a number of writings, bringing up a broad range of notions and topics: movement and change of place, alteration in aspect and form, temporal change, variation in the way a given being is perceived, change in relationship between beings, and qualitative and accidental alterations. He discusses the relationship between cause and effect, the specific characteristics of change (the distinction between real change and merely apparent change), and the relationship between change and the soul. He does not seem to arrive at a unified theory; on the contrary, he appears to adapt his thought to the necessities imposed by individual thematic and theoretical contexts in which one may consider that X passes from state A to state B. Naturally, he applies his own logical and analytical instruments to the various cases, but the different starting points—in different conditions and multiple contexts—frequently determine results that in themselves are perfectly coherent (though at times aporetic), but can only be partially integrated in a unified system. From this point of view, Aristotle's research shows magnificence of conception but also inevitable internal limitation and procedural stiffness. It is also extremely modern, insofar as the problems addressed by the philosopher are still discussed today, and can perhaps be even more satisfactorily examined in our modern terminology, using the tools of modern logic and semantics. Thus, it is simultaneously possible to understand the distance between Aristotle's and the present-day conceptions of becoming and change, and to sense its implications. A good example is provided by the so-called "Cambridge" or "Russellian change," which will be discussed more than once in the following pages.[1]

---

1 I imagine two glasses in front of me (G_1 and G_2); in a given moment (T^1), G_1 is closer to my hand than G_2; if I move in such a way that, in another given moment (T^2), G_2 is closer to my hand, does this imply a substantial change in the relationship between the two glasses, or does the change involve only their relationship with my hand? Does not the fact that their relationship with my hand has changed entail a change in the relationship between the glasses? Clearly, any univocal response to this question is in reality arbitrary, because it inevitably depends on the philosopher's very conception of the world and the historical period in which he or she lives or is read. Present-day responses do not seem to coincide with Aristotle's.

The present volume sets out to provide the reader with some new insights into these subjects. It opens with Robert Wardy's introduction, in which the problem of change is inserted in the context of the great debate surrounding the entire book VII of Aristotle's *Physica*: the causal nexus and mechanism that govern the relationship between cause and effect. Next we give Aristotle's text in Greek according to the so-called version α in the edition by W. D. Ross (Oxford, 1950), which we have modified slightly; this text is accompanied by an English translation. This translation is based on the translations of the six parts of the text proposed by the collaborators and used in their commentaries. Stefano Maso and Gerhard Seel harmonized the terminology of these translations. We also include version β of Aristotle's chapter, which was known by Simplicius, Philoponus, and Themistius. This, too, is given in both the original Greek and an English translation by Stefano Maso and Gerhard Seel. In this way, the reader can easily follow the arguments with reference to all the original materials.

The core of the book contains analyses and commentaries, in two parts: "Preliminary Remarks," by Benjamin Morison and Gerhard Seel, concerning the place of chapter 3 in book VII, the structure of the chapter, and its main problems; and "Commentaries on the Six Sections of the Chapter."

The first section (245b3–246a9), addressed by Ben Morison, gives Aristotle's thesis (alteration is limited to change of perceptible qualities) and shows that change of form cannot count as alteration (ἀλλοίωσις).

Ursula Coope then examines the second section (246a10–b3): neither states of the body (ἕξεις) nor states of the soul are subject to alteration (ἀλλοίωσις), insofar as they are viewed as excellences (ἀρεταί) or defects (κακίαι).

István Bodnár analyzes the third section (246b3–20): Aristotle puts excellences (ἀρεταί) and defects (κακίαι) in the category of relation (τὸ πρός τι πώς ἔχειν) and thus shows that they do not undergo alteration.

In the fourth section (246b20–247a19), studied by Cristina Viano, Aristotle demonstrates the same point for the case of excellences and defects of the soul (τῆς ψυχῆς).

Frans A. J. de Haas examines the fifth section (247b1–13). Here, Aristotle deals with the case of the states of the thinking part of the soul. Their acquisition is not a generation and therefore not an alteration either.

Carlo Natali addresses the sixth and final section (247b13–248a9). Aristotle here presents a further argument in support of his thesis that states of the intellectual part of the soul (ἕξεις) are not the result of generation, nor is the passage from potential knowing to actual knowing (ἠρεμίζεται) a form of alteration.

These six contributions are not simply a reproduction of the 2007 conference papers; on the contrary, they are the result of a process of meticulous revision that additionally took into account both the conference discussions and subsequent correspondence among the participants.

We have added two appendices: Gerhard Seel's "The Logical and Semantic Background of Aristotle's Argument," and Oliver Primavesi's "Aristotle, *Physics* VII.3.245b3–248a9: Towards a Fuller Synopsis of the Two Versions." The latter facilitates the comparison of the two versions (note that Primavesi's text is not identical with the Greek text of versions α and β printed on pp. 18–32). Finally, we give a selective bibliography, which includes texts used explicitly or implicitly by the scholars during the Vitznau meeting.

Introduction

Robert Wardy

Physics VII does not make for an easy read. In the spheres of moral and political philosophy, Aristotle's classic contributions continue to enjoy canonical status, as various virtue ethicists, particularists and communitarians return to his works for inspiration. In the philosophy of mind, keen controversy over Aristotle's psychology does not abate. In metaphysics, essentialist theories, which palpably descend from Aristotelian ontology, are a force to reckon with. Aristotle's moral, political, psychological and metaphysical writings easily engage the attention of contemporary practitioners.

In contrast, his natural philosophy remains relatively inaccessible, the preserve of adepts and *aficionados*. Why is this so? Evidently in the first instance because we regard its central tenets as non-starters: living things are not endowed with irreducible, eternal natures of a teleological character; underlying material constitution is crucially important, and not to be relegated to the subordinate position of a bare enabling condition; the world is not so arranged as to depend for its maintenance on a supernatural Mover . . . And so on, and so forth. Such exotic ideas might captivate the antiquarian imagination, but are rightly confined within the museum of philosophical curiosities.

And things only get worse when one turns to *Phys.* VII. *Phys.* VIII presents a detailed and elaborate set of arguments for the existence of a first Mover: in places they are, of course, intensely problematic, but at least in outline one sees the propositions aimed at clearly enough. VII, however, begins with an intricate argument against kinetic infinity: the logic and, indeed, the precise conclusion of this *reductio* are to a degree opaque. Does the book even hang together? The proof of the first chapter contends that an infinite series of causally dependent changers would be such as to perform an infinite change within a finite time—but that is impossible. To reach this conclusion, Aristotle reasons that since the finite changes lower in the series depend for their occurrence on higher members, all these changes must occur simultaneously, in whatever fixed period is required for completion of the lowest change. He anticipates an objection. Never mind that we have an infinite number of changes: since each single change is nevertheless finite, there is no licit inference to the assumption that together they yield, in aggregate, one infinite change. His response is to postulate that such dependent changers must be in contact, and to assert that contact suffices so to unify the series that it is legitimate to regard it as producing an infinite change—*per impossibile.* Although the *reductio* of the first chapter is formulated in terms of locomotion, Aristotle seems to envision its extension to other kinds of change. Thus the second chapter is devoted to the establishment of the pivotal premise that contact is a universal requirement ranging across Aristotle's kinetic categories. The case for necessary contact in the case of alteration rests on the restriction of such change to perceptible qualities, for which Aristotle is confident the contact condition obtains.

So far the dialectical structure is perspicuous; but this clarity begins to dissipate with chapter three. The third chapter displays great argumentative riches, but a recurrent qualified denial is salient in its overall organization: Aristotle repeatedly says of such-and-such a kind of change that it is not to be identified as alteration—although, perhaps, alteration in the strict sense might be a necessary concomitant of these changes. What is going on? How might the third chapter be related to its predecessors? Reading further only aggravates the puzzle. The fourth chapter is a densely aporetic discussion of the likely illegitimacy of comparisons across heterogeneous kinetic kinds; the fifth formulates mathematical proportionalities which are said to hold between kinetic factors—but not invariably. Furthermore, that the treatise exists for the first three chapters in two versions which are closely related but distinct in both language and argument exacerbates uncertainty about the work's status. A minimal reading would have it that *Phys.* VII is wanting in any real cohesion between its parts, an editorial assemblage of orphaned fragments of Aristotelian natural philosophy. Thus the first two chapters should be coupled as a stunted exercise in proving that there is a First Mover; but the remaining three are bereft of their original contexts, and must accordingly be assessed as free-standing efforts, on whatever merits they might be assigned with regard to intrinsic quality of argumentation and alliance with or repercussions for securely attributed doctrines.

But this style of dismissive reading should only be adopted as a last resort, once the effort to make coherent sense of *Phys.* VII has been abandoned. And there is no reason to give up. True, Aristotle does not tell us how the progression works; but it is not at all unusual for him to leave connections, even difficult ones, inexplicit. Thus we have every reason to speculate about what architectonic rationale might bind the book together, on the understanding that the nature of the case is likely to yield only provisional conclusions and leave a remainder of unresolved opacity. The most probable explanation for the expansion of *Phys.* VII beyond chapters one and two runs as follows. To get chapters 4 and 5 out of the way, since they lie beyond the scope of this book: one might suspect that the *reductio*'s attempt to circumvent the objection runs afoul of Aristotle's own strict criteria for kinetic individuation, which certainly do not countenance contact or continuity as sufficient for the unification of changes. Therefore chapter 4 might best be seen as articulating these concerns. As for chapter 5, its mathematical formulae might perhaps indicate an alternative route around the obstacle. These formulae coordinate dynamic impulse, size of object moved, time and distance travelled; so one might extrapolate that Aristotle could argue that were the series unending, an infinite force would have to be expended during a finite time—but he denies that there is any proportion between finite and infinite.[2] Be that as it may, the function of the third chapter within VII is relatively unobscure. While its predecessor betrays no hesitation over the contact condition's holding good for alteration, the second chapter rests its case on the assertion that the condition is uniformly satisfied for change in perceptible quality. However, a reader conversant with category theory coming to this text might well recollect that Aristotle elsewhere includes diverse subdivisions of the category of quality, and consequently wonder whether and how *Phys.* VII is to accommodate them: do they also satisfy the condition? If not, what are the implications for the *reductio*? So the rationale one most plausibly adduces for inclusion of the third chapter is that it addresses just this concern. As for the existence of two versions, rather than deprecating their existence as a tedious problem, one should welcome the challenge they pose to insightful comparative analysis of the arguments.

A balanced, judicious assessment of such attempts at dialectical contextualization is desirable. On the one hand, looking for the parts to somehow add up is a pretty obvious act of exegetical charity; and the plausible hypothesis that the third chapter fulfills a task whose completion is integral to the *reductio* enjoys a venerable pedigree going back to Alexander.[3] But on the other hand, one should also be careful not to exaggerate the supposed effects of supplying a specific motivation internal to *Phys.* VII. After all, it is not as if validity and soundness are relative to one or another context! However, we should acknowledge that failure to register what Aristotle might be up to on the larger scale can be risky.

Chapter two provides a cautionary example. There Aristotle says "in a way the senses too are altered" (244b10–11). In the longstanding

---

[2] Full exposition of this interpretation is to be found in R. Wardy, *The Chain of Change: A Study of Aristotle's* Physics VII (Cambridge, 1990).

[3] Cited at length and approvingly by Simplicius (*In Aristotelis* Physicorum Libros Quattuor Posteriores *Commentaria*, ed. H. Diels (Berlin, 1895), 1081, 10–30). Cf. 1061, 25–29.

disagreement over the proper construal of his model of perception, much has been made of this statement. So-called "materialists" are keen to emphasize the claim that the senses are altered; their opponents, the "spiritualists," stress the saving qualification, "in a way." Neither side pauses to take account of the fact that this is not an autonomous statement, but rather a premise meant by Aristotle to contribute to verification of the contact condition: the sense organs are, he maintains, in contact with the actualized medium which acts upon them. To forestall misunderstanding, I am of course not pretending that Aristotle would come out with propositions he actually regards as false for the sake of pushing through a convenient argument; nonetheless, what is at the center of his attention can and does have a bearing on whether we evaluate his positions correctly. In the present instance, what must not be lost sight of is that if (something like or less than) alteration is—somehow or other—involved in perception, Aristotle has everything he needs for the purpose in hand, and finer detail would be redundant; ambition to extract decisive evidence about his theory of perception from this indeterminate passage would be imprudent. One wants to be moving back and forth between argumentative structures of smaller and larger scale. A lesson about how to gauge Aristotelian dialectic applicable to the interpretation of *Phys.* VII.3.

To recapitulate: Aristotle defines alteration as change in the category of quality. That category comprises not only perceptible qualities (alternatively, παθητικαὶ ποιότητες, "affective qualities," viz., qualities which are productive of affections in the senses), but also states (ἕξεις), conditions (διαθέσεις), capacities (δυνάμεις), shapes (σχήματα), and external forms (μορφαί) (*Cat.* 8.8b25ff.). VII.2 purports to establish the necessity of contact for change in perceptibles: what about the rest of the category? VII.3 takes care of this unfinished business in a complicated manner: of the remainder Aristotle says that change in them is not alteration, now restricted to change in sensible qualities—but such change does not occur "without alteration."

> One would be most inclined to suppose that alteration occurs both in the possession of shapes, forms and states and in the acquisition and loss of them—but there is no alteration in either case. (245b6–9; cf. 245b21–24)

And who, exactly, would be vulnerable to this rather recondite confusion? The reader of the *Categories*, for one: under the guise of this warning, we catch Aristotle correcting himself. From the narrower perspective of book VII and natural philosophy, what matters is that to suppose that qualities other than the perceptibles are (not) in contact with what creates or affects them is to fall into logical incoherence. Why does this not put paid to the contact condition? Because, first, both the suppositions that they are and that they are not in contact with their changers are illicit, in Rylean terms, "category mistakes"; but second, since they are necessarily associated with alteration—the "not without" clause—perhaps because these changes in perceptibles do necessitate contact, the *reductio* is protected.

From the broader Aristotelian perspective, rescinding from whatever the special project of *Phys.* VII might be, the manœuvre the third chapter executes—"this is not alteration, but does not occur without it"—is positively fascinating. First and simply, the topics canvassed are of huge metaphysical and ethical importance—albeit Aristotle's handling of them can be oblique and problematic. Second and not so simply, the "not without" qualification brings the chapter squarely within the ambit of the grand debates concerning Aristotelian resistance to reductive explanation, and whether he tolerates one or another form of supervenience. At least the majority of affective qualities are logically simple: that is, they are not products of independently specifiable, underlying causal processes, and hence are insusceptible to analytical explanation adverting to more basic changes (Aristotelian red comes about as something reddens, and red really and truly is—red; end of story). The things investigated in VII.3 are logically complex: just what does Aristotle intend, when he insists that changes in them necessarily involve changes in the simple and irreducible? The penetrating contributions which follow are set fair to enhance our understanding on both scores. In the remainder of this introduction I furnish some more orientation and also assemble an aporetic agenda which might usefully be borne in mind as one approaches specific exegetical suggestions.

Shaping and generation (245b9–246a9 / 245b26–246a29). Aristotle relies on a linguistic argument: when something has altered, we designate it by the name borne by the affection (e.g., the heated bronze is called "hot"); but we describe what has been endowed with a shape or generated with a specially adapted term (e.g., the statue is "brazen," not "bronze"). The versions diverge: α infers that linguistic usage shows us that existential change is not alteration (246a1–4), β that alteration occurs exclusively in affective qualities (246a24–25).

Questions.

(i) Demonstrably, shapes and shaping come in for treatment because shapes are qualities, and so one would be forgiven for wrongly thinking that to give something a shape is to alter it. Sometimes what acquires the shape becomes something new (the statue emerges from the bronze): but is that always the case? Complementarily, does the linguistic argument deliver morals about existential change which have application beyond shaping?

(ii) Does Aristotle consider that citation of the linguistic testimony constitutes a probative, or merely suggestive, argument in favour of his conclusion? That is, why should Aristotelian philosophers take notice of "what we say," in this case, the lexical asymmetry? How are we to distinguish between irrelevant, superficial, or even specious linguistic features, and those which might strongly guide or, more modestly, anyway confirm our philosophizing?

(iii) Similar but not identical lexical arguments crop up in *Metaph.* VII.7 and IX.7. What is the significance of this recurrence?

Completion or perfection (246a10–246b3 / 246a25–28). Both versions deny that these are alteration, but use the denial for different purposes. In α, the desired conclusion is that the virtues and vices are not alterations; in β, as an adjunct to the foregoing linguistic argument, that generation is not the same as alteration. In reverse order: β has it that it would be absurd to say that what has been completed has thereby been altered, presumably on the grounds that, if incomplete, it is not yet available for alteration. Question: the point is sound, and integrates without strain into the preceding discussion; so what do we make of the discrepancy between the versions? α has it that neither virtues nor vices are alterations. Virtue is a sort of perfection, vice, a falling away; and if something entirely or most nearly achieves its very nature on perfection, this achievement is not alteration. So this argument too turns on the logic of generation, but also incorporates a teleological view of functional kinds.

Questions.

(i) From here on in, despite the apparent generality of the topic "change in ἕξις," VII.3's treatment focuses on virtuous and vicious ἕξεις. Needless to say, we attend more closely to good and bad conditions than neutral ones; but is it that virtues and vices go proxy for all conditions, or does Aristotle have an implicit argument to the effect that really there are no neutral states, at least when ἕξις has its technical signification?

(ii) Given the link forged between "being an *x*" and "being a good *x*," does this argument import a doctrine of degrees of being? If so, is it an acceptable consequence?

The remainder of VII.3 is a tight sequence, progressing from somatic to ethical to epistemic virtues and vices. There is a constant pattern: if all virtuous and vicious states consist in relations, since relational change is different from, and lesser than, change in substance, quantity or quality, these states will not be subject to qualitative change in particular. How "lesser than"? *Phys.* V.2 enunciates the difference: since relations can come about, be modified or perish while any given *relatum* remains as it was, relational change is not intrinsic to the *relata* (225b11–13)—in the current jargon, this is mere "Cambridge change." Aristotle's formal discussion of relatives, *Categories* chapter 7, is notoriously crabbed, its interpretation, vexed; so one might entertain the hope that VII.3 could shed some light into this darkness.

Somatic ἕξεις are relational (246b3–20 / 246a29–246b27). Again the versions diverge. Excellences and deficiencies of the body are relational inasmuch as we locate them "in" the "mixture" and "harmony" of organic heat and cold (α, 246b5–6), as opposed to a straightforward identification of the somatic ἕξεις with the mixtures (β, 246b21). Good or bad mixture, which affects both internal constitution and relations between the organism and its environment, is responsible for (im)passivity to those affections which tend to preservation or destruction. So α hedges—the virtues and vices are somehow or other consequent upon mixture—while β does not. Question: is the imprecision of α welcome, in that it leaves an opening for a more refined ontology and nuanced biological theorizing, to be pursued elsewhere? Or is this unwelcome vagueness?

Ethical ἕξεις are relational (246b20–247a19 / 246b27–247 a28). The first part of this section simply substitutes social for natural environment, and replicates the preceding argument. The second part, however, which denies that the acquisition and loss of virtues and vices of character are alterations, is distinctive, because its articulation of the necessary accompaniment rider is forthright, unqualifed and unwavering: for such states to be gained or lost, perceptual alteration must occur, period (247a6–7, 247a17–18), immediately followed by the insistence that even necessary association falls short of full-blown identity (247a18–19). How so? Moral character is fixed by experience of and reaction to physical pleasure and pain, themselves either sensations or their upshot. Aristotle argues: virtue and vice concern pleasure and pain, and the former arise when the latter do. But since pleasures and pains are alterations of the sense organs, virtue and vice are necessarily gained and lost on the occurrence of alteration. Therefore their generation is accompanied by, but not identical to, alteration.

Questions.

(i) How can Aristotle allow himself to identify a perceptual act with alteration in a sense organ? As I noted above when discussing my example from the second chapter, there he did not identify them unqualifiedly.

(ii) How can Aristotle identify pleasure and pain with these perceptual alterations? Is this identification a sign that at the time of composition of *Phys.* VII, he adhered to a kinetic theory of pleasure affiliated to various Platonic accounts and incompatible with his celebrated mature theory in *NE* X, which denies that pleasure is a κίνησις?

(iii) Exactly what are the nature and strength of the dependence of the formation of moral character on pleasant and painful experience? When we return to the narrower perspective, we grasp that Aristotle's tactic in this section must be to invariably associate gain and loss of ethical ἕξεις with alteration understood as change in affective quality—but not to identify them. So were pleasure and pain themselves not alterations, the tactic would fail. One might immediately retort that even if provision of a proper, explicit account of pleasure and pain would be excessive and distracting in context, Aristotle could at least have hinted that alteration is not quite all there is to the occurrence of pleasure and pain.

(iv) Are these differences between Aristotle's management of perception and pleasure and pain substantive or no more than ostensible? If the latter, do we explain the appearances away, dismissing them as excusable oversimplification, given that the nature of pleasure and pain as such is peripheral to the job in hand? If the former, what is the genetic story we should tell?

Epistemic ἕξεις are relational (247b1–248a9 / 247a28–248b28).

Not only are the epistemic ἕξεις not alterations, they also do not come about through generation. The potential knower is actualized by entering a relation to an object of knowledge without itself changing intrinsically ("Cambridge change" once again). The versions agree in their confidence that that knowledge is relational is altogether clear; but then perhaps diverge. In β, this relational nature is manifest, in that we derive knowledge of the universal from experience of particulars. In α the text is uncertain. On one reading, it might be in line with β, and says that when confronted by what is particular, the universal is somehow known in or through the particular; on the other reading, contrariwise that the particular is somehow known in or through the universal.

Questions.

(i) How does the genetic argument in β run? *APo.* II.19 inevitably comes to mind, so one imagines that the idea is that coming into the presence of appropriate objects stimulates either concept acquisition or the formulation of certain fundamental propositions (the choice depending on how one reads II.19). But then how could this yield the desired conclusion? For unless the assumed epistemology is a form of radical innatism, coming to be appropriately related to relevant particulars cannot on its own suffice for the acquisition of knowledge. Are there other, superior ways to reconstruct the genetic argument?

(ii) Do we find a different argument in α? Maybe. Perhaps in this version Aristotle has not yet broached the topic of the acquisition of epistemic ἕξεις; rather he is talking about reduction of second potentiality to second actuality, that is, knowers' entering relations with things which fall under and so activate their knowledge. Are there other possibilities?

(iii) Which is the right reading for α? If Aristotle's doctrine that we recognize particulars as tokens of various universal types is in the picture, not much may hang on the choice, since either reading might indifferently express the proposition that no more than relational change is needed to bring particular and universal together in knowledge.

Finally, Aristotle turns to original acquisition of these ἕξεις (that is, if one rejects a genetic reading in α of what precedes). It too is neither generation nor alteration. This is because we achieve ἐπιστήμη when our disorderly minds settle down and our thought comes to rest; but there is no generation of a static state. Aristotle compares acquisition to waking up, or recovery from drunkenness or illness, where we do not regard people as becoming knowledgeable again, despite the temporary unavailability of their knowledge.

Questions.

(i) Is this more than a metaphorical comparison? If not, why be persuaded by Aristotle?

(ii) If it is, is Aristotle committed to innatism? Of what strength?

At the outset I remarked that natural philosophy in general and *Phys.* VII in particular are likely to suffer in comparison with those aspects of Aristotelian metaphysics, ethics and politics which attract the admiring attention of many contemporary philosophers. The natural philosophy can seem quaint, if ingenious; *Phys.* VII might appear especially rebarbative, if one apprehends that the knottiness of the text demands significant mental expenditure for a modest, ambiguous reward. This introductory survey of its third chapter serves to dispel any such unenthusiastic disquiet. The elegant, austere beauties of VII.3's subtle disquisition should easily engross all hardcore dialecticians; thus, at the very least, we might expect that visitors to this room in the Aristotelian wing of the philosophical museum will leave content. But we can and should push further. VII.3 is Aristotle's most sophisticated study of alteration, the category of quality, and how simple qualities might be related to the complex. As the rigorous chapters of this volume amply testify, it just is not true that Aristotle the stimulating metaphysician and ethical theorist is to be separated off from that historical curiosity, Aristotle the antique natural philosopher: the arguments of VII.3 have manifold potential to catalyze reconsideration of key ontological and moral issues. Aristotle is Aristotle, and *Phys.* VII.3 is Aristotle at his challenging best.

Greek Text with English Translation

Harmonized by *Stefano Maso* and *Gerhard Seel*

Version α

Greek text of version α

The most important codices with the α version:

H	=	Vaticanus 1027, saec. xiii or xiv
I	=	Vaticanus 241, saec. xiii
b	=	Parisinus 1859, saec. xiv
c	=	Parisinus 1861, saec. xv
j	=	Parisinus 2033, saec. xv
y	=	Bodleianus Misc. 238, saec. xvi

The codices H and I partially agree with the β version.

Greek Text of Version α

245b3 Ὅτι δὲ τὸ ἀλλοιούμενον ἅπαν ἀλλοιοῦται ὑπὸ τῶν αἰσθητῶν, καὶ ἐν μόνοις ὑπάρχει τούτοις ἀλλοίωσις ὅσα καθ' αὑτὰ λέγεται πάσχειν ὑπὸ τῶν αἰσθητῶν, ἐκ τῶνδε θεωρητέον. τῶν γὰρ ἄλλων μάλιστ' ἄν τις ὑπολάβοι ἔν τε τοῖς σχήμασι καὶ ταῖς μορφαῖς καὶ ἐν ταῖς ἕξεσι καὶ ταῖς τούτων λήψεσι καὶ ἀποβολαῖς ἀλλοίωσιν ὑπάρχειν· ἐν οὐδετέροις δ' ἔστιν. τὸ μὲν γὰρ σχηματιζόμενον καὶ ῥυθμιζόμενον ὅταν ἐπιτελεσθῇ, οὐ λέγομεν ἐκεῖνο ἐξ οὗ ἐστιν, οἷον τὸν ἀνδριάντα χαλκὸν ἢ τὴν πυραμίδα κηρὸν ἢ τὴν κλίνην ξύλον, ἀλλὰ παρωνυμιάζοντες τὸ μὲν χαλκοῦν, τὸ δὲ κήρινον, τὸ δὲ ξύλινον. τὸ δὲ πεπονθὸς καὶ ἠλλοιωμένον προσαγορεύομεν· ὑγρὸν γὰρ καὶ θερμὸν καὶ σκληρὸν τὸν χαλκὸν λέγομεν καὶ τὸν κηρόν (καὶ οὐ μόνον οὕτως, ἀλλὰ καὶ τὸ ὑγρὸν καὶ τὸ θερμὸν χαλκὸν λέγομεν), ὁμωνύμως τῷ πάθει προσαγορεύοντες τὴν
246a1 ὕλην. ὥστ' εἰ κατὰ μὲν τὸ σχῆμα καὶ τὴν μορφὴν οὐ λέγεται τὸ γεγονὸς ἐν ᾧ ἐστι τὸ σχῆμα, κατὰ δὲ τὰ πάθη καὶ τὰς ἀλλοιώσεις λέγεται, φανερὸν ὅτι οὐκ ἂν εἶεν αἱ γενέσεις ἀλλοιώσεις. ἔτι δὲ καὶ εἰπεῖν οὕτως ἄτοπον ἂν δόξειεν, ἠλλοιῶσθαι τὸν ἄνθρωπον ἢ τὴν οἰκίαν ἢ ἄλλο ὁτιοῦν τῶν γεγενημένων· ἀλλὰ γίγνεσθαι μὲν ἴσως ἕκαστον ἀναγκαῖον ἀλλοιουμένου τινός, οἷον τῆς ὕλης πυκνουμένης ἢ μανουμένης ἢ θερμαινομένης ἢ ψυχομένης, οὐ μέντοι τὰ γιγνόμενά γε ἀλλοιοῦται, οὐδ' ἡ γένεσις αὐτῶν ἀλλοίωσίς ἐστιν.

ἀλλὰ μὴν οὐδ' αἱ ἕξεις οὔθ' αἱ τοῦ σώματος οὔθ' αἱ τῆς ψυχῆς ἀλλοιώσεις. αἱ μὲν γὰρ ἀρεταὶ αἱ δὲ κακίαι τῶν ἕξεων· οὐκ ἔστι δὲ οὔτε ἡ ἀρετὴ οὔτε ἡ κακία ἀλλοίωσις, ἀλλ' ἡ μὲν ἀρετὴ τελείωσίς τις (ὅταν γὰρ λάβῃ τὴν αὑτοῦ ἀρετήν, τότε λέγεται τέλειον ἕκαστον – τότε γὰρ ἔστι μάλιστα [τὸ] κατὰ φύσιν – ὥσπερ κύκλος τέλειος, ὅταν μάλιστα γένηται κύκλος καὶ ὅταν βέλτιστος), ἡ δὲ κακία φθορὰ τούτου

246a15: [τὸ] Ross

English Translation of Version α

That everything which is altered is altered by perceptible quali- **245**b3
ties, that is, that alteration takes place only in those things which are said to be affected in their own right by perceptible qualities, is to be seen from the following considerations. For of the other [cases], someone would most of all suppose that alteration takes place in figures and shapes, and in states, both in the taking on and casting off of them: but in neither of them [is there alteration]. For we do not say that that which is being shaped and fashioned, once it has been completed, is that which from which it is made. For instance, we do not say that the statue is bronze, or the candle is wax, or the bed wood, but using paronyms, we say that they are brazen, waxen, and wooden. But we do call that which has been affected and altered [that from which it is altered]: for we call the bronze fluid and hot and hard, and the wax too, and not only this way round, but also we call that which is fluid, and that which is hot, bronze, calling the matter homonymously with the affection.
The result is that, in the case of figure and shape, that which has **246**a1
come to be is not called the thing in which the shape is, but in the case of affections and alterations, [that which has come to be] is called [the thing in which the affection and the alteration is], it is clear that these instances of becoming would not be alterations. Again, it would seem to be absurd also to say that the man had been altered, or the house, or many other things which have come to be; but perhaps it is necessary that each comes to be with something being altered, e.g., the matter being condensed or rarefied, or heated or cooled, but those things which come to be are not being altered, nor is the coming to be of them an alteration.

Nor indeed are states, either of the body or of the soul, alterations. For some of them are excellences and others are defects, but neither excellence nor defect is an alteration. On the contrary, excellence is a kind of perfection (for in each case, when something has acquired its own excellence, then it is called perfect—for then it is most in accordance with [its own] nature—just as a circle is perfect, when it has become a circle in the highest degree and

καὶ ἔκστασις· ὥσπερ οὖν οὐδὲ τὸ τῆς οἰκίας τελείωμα λέγομεν ἀλλοίωσιν (ἄτοπον γὰρ εἰ ὁ θριγκὸς καὶ ὁ κέραμος ἀλλοίωσις, ἢ εἰ θριγκουμένη καὶ κεραμουμένη ἀλλοιοῦται ἀλλὰ μὴ τελειοῦται ἡ οἰκία), τὸν αὐτὸν τρόπον καὶ ἐπὶ τῶν ἀρε- **246b1** τῶν καὶ τῶν κακιῶν καὶ τῶν ἐχόντων ἢ λαμβανόντων· αἱ μὲν γὰρ τελειώσεις αἱ δὲ ἐκστάσεις εἰσίν, ὥστ' οὐκ ἀλλοιώσεις.

ἔτι δὲ καί φαμεν ἁπάσας εἶναι τὰς ἀρετὰς ἐν τῷ πρός τι πὼς ἔχειν. τὰς μὲν γὰρ τοῦ σώματος, οἷον ὑγίειαν καὶ εὐεξίαν, ἐν κράσει καὶ συμμετρίᾳ θερμῶν καὶ ψυχρῶν τίθεμεν, ἢ αὐτῶν πρὸς αὑτὰ τῶν ἐντὸς ἢ πρὸς τὸ περιέχον· ὁμοίως δὲ καὶ τὸ κάλλος καὶ τὴν ἰσχὺν καὶ τὰς ἄλλας ἀρετὰς καὶ κακίας. ἑκάστη γάρ ἐστι < ἐν > τῷ πρός τι πὼς ἔχειν, καὶ περὶ τὰ οἰκεῖα πάθη εὖ ἢ κακῶς διατίθησι τὸ ἔχον· οἰκεῖα δ' ὑφ' ὧν γίγνεσθαι καὶ φθείρεσθαι πέφυκεν. ἐπεὶ οὖν τὰ πρός τι οὔτε αὐτά ἐστιν ἀλλοιώσεις, οὔτε ἔστιν αὐτῶν ἀλλοίωσις οὐδὲ γένεσις οὐδ' ὅλως μεταβολὴ οὐδεμία, φανερὸν ὅτι οὔθ' αἱ ἕξεις οὔθ' αἱ τῶν ἕξεων ἀποβολαὶ καὶ λήψεις ἀλλοιώσεις εἰσίν, ἀλλὰ γίγνεσθαι μὲν ἴσως αὐτὰς καὶ φθείρεσθαι ἀλλοιουμένων τινῶν ἀνάγκη, καθάπερ καὶ τὸ εἶδος καὶ τὴν μορφήν, οἷον θερμῶν καὶ ψυχρῶν ἢ ξηρῶν καὶ ὑγρῶν, ἢ ἐν οἷς τυγχάνουσιν οὖσαι πρώτοις. περὶ ταῦτα γὰρ ἑκάστη λέγεται κακία καὶ ἀρετή, ὑφ' ὧν ἀλλοιοῦσθαι πέφυκε τὸ ἔχον· ἡ μὲν γὰρ ἀρετὴ ποιεῖ ἢ ἀπαθὲς ἢ ὡδὶ παθητικόν, ἡ δὲ κακία παθητικὸν ἢ ἐναντίως ἀπαθές.

246b8: <ἐν> Simplicius

[has become] best), and defect, on the other hand, is destruction of and departure from this. So just as we do not call the perfection of the house an alteration (for it would be strange if the coping and tiling were called alteration, or if, in being coped and tiled, the house were altered instead of being perfected), in the same way [do we speak] also of excellences and defects, and of the things that **246b1** possess or acquire them. For the first [the excellences] are perfections and the latter [the defects] are departures, and consequently neither are alterations.

Moreover, we say that all the excellences are [in the category of being] related in some way to something. For we place the excellences of the body—like health or fitness—in the mixture and due proportion of warm and cold things—either [in the mixture and due proportion] of the internal components in relation to themselves, or [in the mixture and due proportions] in relation to the surrounding elements. [We place] in a similar manner also beauty, and strength and the other excellences and defects. For each of them is [in the category of] being related in some way to something, and they put that which has them in a good or bad condition in relation to its proper affections—that is those by which it comes into being or perishes according to its nature. Since, then, the relations are neither alterations themselves, nor is there an alteration of them, nor [is there] generation, or any change whatsoever [of them], it is clear that neither the states, nor the loss or the acquisition of states are alterations, but perhaps it is necessary that they come into being and perish when some [other] things undergo alteration, as [it is necessary that] shape and form [is generated when] e.g., warm and cold and wet and dry things [undergo alteration], or those things in which [shape and form] happen to be as in the first things [which they are in]. For each defect or excellence is said in regard of those [affections] by which that which has [the defect or the excellence] undergoes alteration according to its nature. For the excellence makes [that which has it] either liable not to be affected, or to be affected in a particular way, whereas defect [makes it] liable to be affected, or [makes it] liable not to be affected in the opposite way.

246b8 We read ἐν τῷ with Simplicius

ὁμοίως

247a1 δὲ καὶ ἐπὶ τῶν τῆς ψυχῆς ἕξεων· ἅπασαι γὰρ καὶ αὗται τῷ πρός τι πὼς ἔχειν, καὶ αἱ μὲν ἀρεταὶ τελειώσεις, αἱ δὲ κακίαι ἐκστάσεις. ἔτι δὲ ἡ μὲν ἀρετὴ εὖ διατίθησι πρὸς τὰ οἰκεῖα πάθη, ἡ δὲ κακία κακῶς. ὥστ' οὐδ' αὗται ἔσονται ἀλλοιώσεις· οὐδὲ δὴ αἱ ἀποβολαὶ καὶ αἱ λήψεις αὐτῶν. γίγνεσθαι δ' αὐτὰς ἀναγκαῖον ἀλλοιουμένου τοῦ αἰσθητικοῦ μέρους. ἀλλοιωθήσεται δ' ὑπὸ τῶν αἰσθητῶν· ἅπασα γὰρ ἡ ἠθικὴ ἀρετὴ περὶ ἡδονὰς καὶ λύπας τὰς σωματικάς, αὗται δὲ ἢ ἐν τῷ πράττειν ἢ ἐν τῷ μεμνῆσθαι ἢ ἐν τῷ ἐλπίζειν. αἱ μὲν οὖν ἐν τῇ πράξει κατὰ τὴν αἴσθησίν εἰσιν, ὥσθ' ὑπ' αἰσθητοῦ τινὸς κινεῖσθαι, αἱ δ' ἐν τῇ μνήμῃ καὶ ἐν τῇ ἐλπίδι ἀπὸ ταύτης εἰσίν· ἢ γὰρ οἷα ἔπαθον μεμνημένοι ἥδονται, ἢ ἐλπίζοντες οἷα μέλλουσιν. ὥστ' ἀνάγκη πᾶσαν τὴν τοιαύτην ἡδονὴν ὑπὸ τῶν αἰσθητῶν γίγνεσθαι. ἐπεὶ δ' ἡδονῆς καὶ λύπης ἐγγιγνομένης καὶ ἡ κακία καὶ ἡ ἀρετὴ ἐγγίγνεται (περὶ ταύτας γάρ εἰσιν), αἱ δ' ἡδοναὶ καὶ αἱ λῦπαι ἀλλοιώσεις τοῦ αἰσθητικοῦ, φανερὸν ὅτι ἀλλοιουμένου τινὸς ἀνάγκη καὶ ταύτας ἀποβάλλειν καὶ λαμβάνειν. ὥσθ' ἡ μὲν γένεσις αὐτῶν μετ' ἀλλοιώσεως, αὐταὶ δ' οὐκ εἰσὶν ἀλλοιώσεις.

247b1 ἀλλὰ μὴν οὐδ' αἱ τοῦ νοητικοῦ μέρους ἕξεις ἀλλοιώσεις, οὐδ' ἔστιν αὐτῶν γένεσις. πολὺ γὰρ μάλιστα τὸ ἐπιστῆμον ἐν τῷ πρός τι πὼς ἔχειν λέγομεν. ἔτι δὲ καὶ φανερὸν ὅτι οὐκ ἔστιν αὐτῶν γένεσις· τὸ γὰρ κατὰ δύναμιν ἐπιστῆμον οὐδὲν αὐτὸ κινηθὲν ἀλλὰ τῷ ἄλλο ὑπάρξαι γίγνεται ἐπιστῆμον. ὅταν γὰρ γένηται τὸ κατὰ μέρος, ἐπίσταταί πως τὰ καθόλου τῷ

The same is the case concerning the states of the soul. All of them consist 247a1
in some particular relation to something, and the excellences are perfections and the defects are departures from perfection. Moreover the excellence puts its owner into a good disposition, with regard to its own affections, while defect puts him in a bad disposition. Consequently these will not be alterations, nor will the loss or acquisition of them be such. But it is necessary that they come to be when the sensible part [of the soul] is altered. And this is altered by sensible things. For all the ethical virtues are concerned with bodily pleasures and pains. But these occur either in action or in memory or in expectation. Now those which occur in action are an effect of sense perception, consequently they are set in motion by something sensible; on the other hand, those which occur in memory or in expectation come from this (for people feel pleasure either in remembering what they have experienced or in expecting what may happen in the future). Thus it is necessary that all pleasure of this kind is produced by sensible things. Since when pleasure and pain is generated [in the soul], defect or excellence is generated as well (for they concern the former) and since pleasures and pains are alterations of the sensible [part of the soul], it is evident that when something is altered necessarily also these states are lost or acquired. Thus, though their coming to be is accompanied by an alteration, they themselves are not alterations.

However, the states of the intellectual part [of the soul] are 247b1
not alterations either, nor is there any generation of them. For we much rather say that that which knows is in [the category of] being related somehow to something. And further, it is evident that there is no generation of them. For that which potentially knows becomes that which [actually] knows not because it has changed itself, but because something else is present. For when

ἐν μέρει. πάλιν δὲ τῆς χρήσεως καὶ τῆς ἐνεργείας οὐκ ἔστι γένεσις, εἰ μή τις καὶ τῆς ἀναβλέψεως καὶ τῆς ἀφῆς οἴεται γένεσιν εἶναι· τὸ γὰρ χρῆσθαι καὶ τὸ ἐνεργεῖν ὅμοιον τούτοις. ἡ δ' ἐξ ἀρχῆς λῆψις τῆς ἐπιστήμης γένεσις οὐκ ἔστιν οὐδ' ἀλλοίωσις· τῷ γὰρ ἠρεμῆσαι καὶ στῆναι τὴν διάνοιαν ἐπίστασθαι καὶ φρονεῖν λεγόμεθα, εἰς δὲ τὸ ἠρεμεῖν οὐκ ἔστι γένεσις· ὅλως γὰρ οὐδεμιᾶς μεταβολῆς, καθάπερ εἴρηται πρότερον.

ἔτι δ' ὥσπερ ὅταν ἐκ τοῦ μεθύειν ἢ καθεύδειν ἢ νοσεῖν εἰς τἀναντία μεταστῇ τις, οὔ φαμεν ἐπιστήμονα γεγονέναι πάλιν (καίτοι ἀδύνατος ἦν τῇ ἐπιστήμῃ χρῆσθαι πρότερον), οὕτως οὐδ' ὅταν ἐξ ἀρχῆς λαμβάνῃ τὴν ἕξιν· τῷ γὰρ καθίστασθαι τὴν ψυχὴν ἐκ τῆς φυσικῆς ταραχῆς φρόνιμόν τι γίγνεται καὶ ἐπιστῆμον. διὸ καὶ τὰ παιδία οὔτε μανθάνειν δύνανται οὔτε κατὰ τὰς αἰσθήσεις ὁμοίως κρίνειν τοῖς πρεσβυτέροις· πολλὴ γὰρ ἡ ταραχὴ καὶ ἡ κίνησις. καθίσταται δὲ καὶ ἠρεμίζεται πρὸς ἔνια μὲν ὑπὸ τῆς φύσεως αὐτῆς, πρὸς ἔνια δ' ὑπ' ἄλλων, ἐν ἀμφοτέροις δὲ ἀλλοιουμένων τινῶν τῶν ἐν τῷ σώματι, καθάπερ ἐπὶ τῆς χρήσεως καὶ τῆς ἐνεργείας, ὅταν νήφων γένηται καὶ ἐγερθῇ. φανερὸν οὖν ἐκ τῶν εἰρημένων ὅτι τὸ ἀλλοιοῦσθαι καὶ ἡ ἀλλοίωσις ἔν τε τοῖς αἰσθητοῖς γίγνεται καὶ ἐν τῷ αἰσθητικῷ μορίῳ τῆς ψυχῆς, ἐν ἄλλῳ δ' οὐδενὶ πλὴν κατὰ συμβεβηκός.

the particular comes to be, [that which knows] knows in a way the universal by means of the particular. Again, there is no generation of the use and actuality [of the state], unless one believes that there is generation of opening one's eyes and touching, for using and being in actuality are similar to these. The original grasp of knowledge is not a generation nor an alteration. For we are said to know and to have understanding when our thinking has come to rest and has reached a standstill, but there is no generation that leads to being in a state of rest; indeed, in general [there is no generation] of any change, as was said before.

Furthermore, just as when someone has passed from drunkenness or sleep or illness to the opposite condition, we do not assert that he has reacquired knowledge by a process of generation, despite his having previously been incapable of using his knowledge, so neither [does this occur] when someone acquires the state in the first place. For it is by the soul's coming to rest from its natural disturbance that someone comes to be wise and knowledgeable (that is also why children are incapable of either learning or discriminating by means of the senses as well as their elders, i.e., because the disturbance and change [during childhood] are great) and it settles down and comes to rest, with regard to some objects because of nature itself, with regard to others under other influences. In either case the processes are accompanied by the alteration of certain things in the body, just as in the transition to use and activity, when someone becomes sober and wakes up. Thus it is evident from what has being said that being altered and alteration occur in perceptible qualities and in the sensitive part of the soul, and in nothing else except *per accidens*.

Version β

Greek text of version β

The most important codices with the β version:

E = Parisinus gr. 1853, beginning of saec. x
F = Laurentianus 87.7, saec. xiv
J = Vindobonensis 100 (olim 34), saec. x
K = Laurentianus 87.24, saec. xiii

Greek text of Version β

245b19 Ὅτι δὲ τὰ ἀλλοιούμενα ἀλλοιοῦνται πάντα ὑπὸ τῶν αἰσθητῶν, καὶ
μόνων τούτων ἔστιν ἀλλοίωσις ὅσα καθ' αὑτὰ πάσχει ὑπὸ τούτων,
21 ἐκ τῶνδε θεωρήσωμεν. τῶν γὰρ ἄλλων μάλιστα [ἄν τις ὑπολάβοι ἔν τε] τοῖς
σχήμασι καὶ ταῖς μορφαῖς καὶ ἐν ταῖς ἕξεσι καὶ ταῖς τούτων ἀποβολαῖς
23 καὶ λήψεσιν [ἀλλοίωσιν ὑπάρχειν.] δοκεῖ [γὰρ] ὑπάρχειν τὸ τῆς ἀλλοιώσεως,
οὐκ ἔστιν δὲ οὐδ' ἐν τούτοις, ἀλλὰ γίγνεται [τὸ σχῆμα] ἀλλοιουμένων τινῶν
25 ταῦτα (πυκνουμένης γὰρ ἢ μανουμένης ἢ θερμαινομένης ἢ ψυχομένης τῆς
ὕλης), ἀλλοίωσις δὲ οὐκ ἔστιν. ἐξ οὗ μὲν γὰρ ἡ μορφὴ τοῦ ἀνδριάντος, οὐ λέ-
27 γομεν τὴν μορφήν, οὐδ' ἐξ οὗ τὸ σχῆμα τῆς πυραμίδος ἢ τῆς κλίνης, ἀλλὰ
παρωνυμιάζοντες τὸ μὲν χαλκοῦν τὸ δὲ κήρινον τὸ δὲ ξύλινον.

τὸ δ' ἀλ-
29 λοιούμενον λέγομεν · τὸν γὰρ χαλκὸν ὑγρὸν εἶναι λέγομεν ἢ θερμὸν ἢ σκλη-
246a21 ρόν (καὶ οὐ μόνον οὕτως, ἀλλὰ καὶ τὸ ὑγρὸν καὶ τὸ θερμὸν χαλκόν),
ὁμωνύμως λέγοντες τῷ πάθει τὴν ὕλην. ἐπεὶ οὖν ἐξ οὗ μὲν ἡ μορφὴ καὶ
23 τὸ σχῆμα καὶ τὸ γεγονὸς ὁμωνύμως οὐ λέγεται τοῖς ἐξ ἐκείνου σχήμασιν,
τὸ δ 'ἀλλοιούμενον τοῖς πάθεσιν ὁμωνύμως λέγεται, φανερὸν ὡς ἐν μόνοις
25 τοῖς αἰσθητοῖς ἡ ἀλλοίωσις.

ἔτι καὶ ἄλλως ἄτοπον. τὸ γὰρ λέγει τὸν
ἄνθρωπον ἠλλοιῶσθαι ἢ τὴν οἰκίαν λαβοῦσαν τέλος γελοῖον, εἰ τὴν τελείω-
27 σιν τῆς οἰκίας, τον' θριγκὸν ἢ τὴν κεραμίδα, φήσομην ἀλλοίωσιν εἶναι, <ἢ> θριγ-
κουμένης τῆς οἰκίας ἢ κεραμιδουμένης ἀλλοιοῦσθαι τὴν οἰκίαν, δῆλον δὴ
29 ὅτι τὸ τῆς ἀλλοιώσεως οὐκ ἔστιν ἐν τοῖς γιγνομένοις.

245b21 [ἄν τις ὑπολάβοι ἔν τε] om. EI[1]
245b23 [ἀλλοίωσιν ὑπάρχειν.] om. EFI[1]JK [γὰρ] om. EI[1]
245b24 [τὸ σχῆμα] om. EFJK

English Translation of Version β

That all things which are altered are altered by perceptible **245**b19
qualities, and that there is alteration only of those things which
are affected in their own right by these [qualities] we will see from 21
the following considerations. For of the other cases, it seems that
[the phenomenon of] alteration belongs mostly to shapes and forms,
and states, both in the casting off and taking on of them, but it 23
[alteration] does not exist in these either. Rather these [shapes and
forms] come to be when something is altered; (I mean, when the 25
matter is condensed or rarefied or heated or cooled); but this [the
former] is not alteration. For, we do not call the [produced] form [by
the name of] the matter out of which the form of the statue was
shaped, nor [by the names of the matter] out of which the shape of 27
the candle or of the bed were made, rather by using a paronym we
say that the first is brazen, the other waxen and the third wooden,
while we call [by these kinds of names] the things that undergo 29
alterations. For we say that the bronze is fluid or hot or hard (and **246**a21
not only in this way, but [we say] also that the fluid and the hot are
bronze), attributing in a homonymous way the matter to the affec-
tion. Now, because [the name of the matter] out of which the form
and the shape and [generally] the generated thing was made is not 23
in a homonymous way attributed to the forms produced out of the
former, while [the name of] the matter that undergoes an alteration
is in a homonymous way attributed to the affections, it is evident
that alteration occurs only to perceptible qualities. Furthermore, it 25
would be absurd in another way too. For it would be ridiculous to
say that the man or the house that has reached perfection has been 27
altered if we say that the perfection of the house, i.e., the coping or
the tiling, is an alteration, or that the house is altered when it is
coped or tiled. Thus it is evident that the [phenomenon] of altera- 29
tion does not exist in things coming to be.

245b21 *it seems . . . belongs mostly*: we follow codices EI
245b23 *taking on of them*: we follow codices EFIJK
245b24 *Rather these*: we follow codices EFJK

οὐδὲ γὰρ ἐν ταῖς ἕξε-
σιν. αἱ γὰρ ἕξεις ἀρεταὶ καὶ κακίαι, ἀρετὴ δὲ πᾶσα καὶ κακία τῶν
246[b]21 πρός τι, καθάπερ ἡ μὲν ὑγίεια θερμῶν καὶ ψυχρῶν συμμετρία τις, ἢ τῶν
ἐντὸς ἢ πρὸς τὸ περιέχον. ὁμοίως δὲ καὶ τὸ κάλλος καὶ ἡ ἰσχὺς τῶν
23 πρός τι. διαθέσεις γάρ τινες τοῦ βελτίστου πρὸς τὸ ἄριστον, λέγω δὲ τὸ
βέλτιστον τὸ σῶζον καὶ διατιθὲν περὶ τὴν φύσιν. ἐπεὶ οὖν αἱ μὲν ἀρεταὶ
25 καὶ αἱ κακίαι τῶν πρός τι, ταῦτα δὲ οὔτε γενέσεις εἰσὶν οὔτε γένεσις αὐ-
τῶν οὐδ᾽ ὅλως ἀλλοίωσις, φανερὸν ὡς οὐκ ἔστιν ὅλως τὸ τῆς ἀλλοιώσεως
27 περὶ τὰς ἕξεις.

οὐδὲ δὴ περὶ τὰς τῆς ψυχῆς ἀρετὰς καὶ κακίας. ἡ μὲν
γὰρ ἀρετὴ τελείωσίς τις (ἕκαστον γὰρ τότε μάλιστα τέλειόν ἐστιν, ὅταν
τύχῃ τῆς οἰκείας ἀρετῆς, καὶ μάλιστα κατὰ φύσιν, καθάπερ ὁ κύκλος
29 τότε μάλιστα κατὰ φύσιν ἐστίν, ὅταν μάλιστα κύκλος ᾖ), ἡ δὲ κακία
247[a]20 φθορὰ τούτων καὶ ἔκστασις. γίγνεται μὲν οὖν ἀλλοιουμένου τινὸς ἡ λῆψις
τῆς ἀρετῆς καὶ ἡ τῆς κακίας ἀποβολή, ἀλλοίωσις μέντοι τούτων οὐδέτερον.
ὅτι δ᾽ ἀλλοιοῦταί τι, δῆλον. ἡ μὲν γὰρ ἀρετὴ ἤτοι ἀπάθειά τις ἢ παθη-
τικὸν ὡδί, ἡ δὲ κακία παθητικὸν ἢ ἐναντία πάθησις τῇ ἀρετῇ. καὶ τὸ
23 ὅλον τὴν ἠθικὴν ἀρετὴν ἐν ἡδοναῖς καὶ λύπαις εἶναι συμβέβηκεν· ἢ γὰρ
κατ᾽ ἐνέργειαν τὸ τῆς ἡδονῆς ἢ διὰ μνήμην ἢ ἀπὸ τῆς ἐλπίδος. εἰ μὲν οὖν
25 κατ᾽ ἐνέργειαν, αἴσθησις τὸ αἴτιον, εἰ δὲ διὰ μνήμην ἢ δι᾽ ἐλπίδα, ἀπὸ
ταύτης· ἢ γὰρ οἷα ἐπάθομεν μεμνημένοις τὸ τῆς ἡδονῆς ἢ οἷα πεισόμεθα
27 ἐλπίζουσιν.

Nor does it exist in states.
For states are excellences and defects, but every excellence and
defect is in [the category of] relation, just as health is the right **246**b21
proportion of hot and cold elements either of those inside or [of
the latter] in relation with the surrounding elements. In a similar
way beauty and strength are in [the category of] relation. For they 23
are certain dispositions of the best [part of body or soul] toward
the highest good, and I call "the best" what preserves and what
is arranged according to [its] nature. Since, then, both excellences
and defects are [in the category] of relation and neither are these 25
generations, nor is there a generation of them or any alteration at
all, it is clear that the [phenomenon] of alteration does not concern 27
states at all. It [the phenomenon of alteration] does not concern the
excellences and defects of the soul either. For excellence is a kind
of perfection (for each thing is most perfect at the very moment,
when it reaches its own excellence and is most in accordance with 29
its own nature, just as a circular body is most in accordance with
its nature when it is most a circle), defect on the other hand is **247**a20
destruction of and departure from these [perfections].

To be sure, both the acquisition
of excellence and the loss of defect take place because something 21
[else] is altered, however none of these [processes] is alteration.
However, it is evident that [in these cases] something is altered.
For excellence is a kind of immunity to affection or a capacity to
be affected in a certain way and defect is the capacity to suffer 23
affection or the affection opposed to excellence. And the totality of ethical excellence happens to exist in pleasures and pains.
For, the [phenomenon] of pleasure occurs either in actuality or 25
through memory or by hope. If it occurs in actuality the cause
[of it] is sense perception, if [it occurs] by memory or hope, it [is
caused] by the same. For the [phenomenon] of pleasure occurs 27
either when we remember what we experienced or when we hope
to experience this in the future.

ἀλλὰ μὴν οὐδὲ τῷ διανοητικῷ μέρει τῆς ψυχῆς ἀλλοίωσις.
29 τὸ γὰρ ἐπιστῆμον μάλιστα τῶν πρός τι λέγεται. τοῦτο δὲ δῆλον· κατ' οὐδε-
μίαν γὰρ δύναμιν κινηθεῖσιν ἐγγίγνεται τὸ τῆς ἐπιστήμης, ἀλλ' ὑπάρξαντός
247^{b}20 τινος· ἐκ γὰρ τῆς κατὰ μέρος ἐμπειρίας τὴν καθόλου λαμβάνομεν ἐπιστή-
μην. οὐδὲ δὴ ἡ ἐνέργεια γένεσις, εἰ μή τις καὶ τὴν ἀνάβλεψιν καὶ τὴν
ἁφὴν γενέσεις φησίν· τοιοῦτον γὰρ ἡ ἐνέργεια. ἡ δὲ ἐξ ἀρχῆς λῆψις τῆς
23 ἐπιστήμης οὐκ ἔστι γένεσις οὐδ' ἀλλοίωσις· τῷ γὰρ ἠρεμίζεσθαι καὶ καθ-
ίστασθαι τὴν ψυχὴν ἐπιστήμων γίγνεται καὶ φρόνιμος.

καθάπερ οὖν οὐδ' ὅταν
καθεύδων ἐγερθῇ τις ἢ μεθύων παύσηται ἢ νοσῶν καταστῇ, γέγονεν ἐπι-
στήμων· καίτοι πρότερον οὐκ ἐδύνατο χρῆσθαι καὶ κατὰ τὴν ἐπιστήμην ἐνερ-
25 γεῖν, εἶτα ἀπαλλαγείσης τῆς ταραχῆς καὶ εἰς ἠρεμίαν καὶ κατάστασιν ἐλθούσης τῆς
διανοίας ὑπῆρξεν ἡ δύναμις ἡ πρὸς τὴν τῆς ἐπιστήμης χρείαν. τοιοῦτο δή
27 τι γίγνεται καὶ τὸ ἐξ ἀρχῆς ἐν τῇ τῆς ἐπιστήμης ὑπαρχῇ· τῆς γὰρ ταρα-
χῆς ἠρεμία τις καὶ κατάστασις. οὐδὲ δὴ τὰ παιδία δύναται μαθεῖν οὐδὲ
29 κρίνειν ταῖς αἰσθήσεσιν ὁμοίως τοῖς πρεσβυτέροις. πολλὴ γὰρ ἡ ταραχὴ
περὶ ταῦτα καὶ ἡ κίνησις. καθίσταται δὲ καὶ παύεται τῆς ταραχῆς τοτὲ
248^{a}26 μὲν ὑπὸ τῆς φύσεως τοτὲ δ' ὑπ' ἄλλων. ἐν ἀμφοτέροις δὲ τούτοις ἀλ-
27 λοιοῦσθαί τι συμβαίνει, καθάπερ ὅταν ἐγερθῇ καὶ γένηται νήφων πρὸς τὴν
ἐνέργειαν. φανερὸν οὖν ὅτι τὸ τῆς ἀλλοιώσεως ἐν τοῖς αἰσθητοῖς καὶ ἐν τῷ
248^{b}26 αἰσθητικῷ μέρει τῆς ψυχῆς, ἐν ἄλλῳ δ' οὐθενὶ πλὴν κατὰ συμβεβηκός.
27

247^{a}28 ἀλλὰ μὴν οὐδ' <ἐν> τῷ add. Spengel, Ross

However, no alteration belongs to the intellectual part of the
soul either. For that which knows is mostly said in the [category] 29
of relation. This is clear from the fact that the [phenomenon] of
knowledge does not arise in those who have undergone change
according to some potentiality, but only if something [else] is
present. For we acquire knowledge of the universal from the **247b20**
experience of particulars.

Nor is the activity [of knowledge] a kind of generation, unless
someone were to call the opening of one's eyes or the act of touch-
ing a generation. For the activity is of this kind. However, the
original acquisition of knowledge is neither a coming to be nor 23
an alteration. For when the soul comes to a rest and reaches a
standstill, someone becomes a knowledgable and wise man.

Likewise if
someone who wakes up from sleeping or gets sober after having 25
been drunk or recovers from sickness [and then starts exercising
his knowledge] did not become knowing despite the fact that
before he was not able to use his knowledge and to exercise it, but
as soon as his disturbance has been changed and his mind has 27
come to rest and reached a standstill, the capacity of using his
knowledge belongs to him [again]. Something like this happens 29
also when knowledge is first acquired, namely a certain calming
down and coming to rest of the disturbance. Children are not
able to learn or to discriminate by sense perceptions to the same **248a26**
degree as adults. For the amount of disturbance and change in 27
regard of these [learning and perceptual discrimination] is huge.
But the disturbance settles down and comes to rest sometimes by
nature sometimes by other influences. In both of these cases it
happens that something is altered, like when someone wakes up **248b26**
and gets sober in regard of the exercise [of his capacity]. Thus it 27
is clear that the [phenomenon] of alteration occurs in perceptible
qualities and in the sensible part of the soul, and in nothing else
except *per accidens*.

Analysis and Commentary

Preliminary Remarks

Benjamin Morison and *Gerhard Seel*

The Place of Chapter 3 in Book VII

Book VII does not belong with books V, VI, and VIII, and Eudemus did not have a book corresponding to it in his *Physics*. Still, it clearly shares subject matter with the other books, and may well have borne the same title: Περὶ κινήσεως. Its central claim in chapter 1 is a key axiom for book VIII, which presumably explains why some editor decided to put it before book VIII, and it contains back references to book V, which is why it was put after that book; books V and VI go so closely together that no one would think of dissociating them.

The other striking anomaly about it is that the first three chapters exist in two versions, both transmitted as the first chapters of book VII. This is unique for Aristotle. As far as we can see, no one has yet produced a good explanation for it. We share Ross's view that the β version is inferior, although Wardy is more circumspect (see his introduction to this volume).

Chapter 1: An argument for the principle that everything that moves is moved by something, and for the claim that such a series of movers cannot be infinite.

Chapter 2: An argument (by cases) that motion involves contact, i.e., that the mover and the moved are always in contact. Aristotle takes local motion, alteration, and increase/decrease in turn.

Chapters 4 and 5 deal with other issues arising from chapter 1, in particular how we are to compare movements to each other. (A key presupposition of the argument in chapter 1 is that we can do so).

Chapter 3 examines the case of alteration more closely, and shows or attempts to show that certain changes that might have looked like alterations are not in fact alterations (see Wardy's introduction to this volume).

The Structure of VII.3

In 245b3–5 Aristotle formulates two principles concerning alteration: 1. Everything which is altered is altered by perceptible qualities. 2. Alteration takes place only in those things which are said to be affected in their own right by perceptible qualities. In b5–6 he announces a demonstration or defence of these principles, which is in fact given in the rest of the chapter. According to the second principle this demonstration has to proceed by way of exclusion, i.e., it has to show that there are no cases of alteration other than those where something is altered through being affected by a perceptible quality in its own right.

In 245b6–9 he mentions as the most likely other cases: alteration in shapes, forms and states and in the taking on and casting off of them. He adds that neither of them is alteration.

According to this disposition he first considers the case of shapes and forms (245b9–246a9), taking the case of coming to be as an essential feature of these (246a1–9). After this he has to investigate the case of states and the acquiring and loosing thereof. As he distinguishes states of the body and states of the soul he has to consider both. He first gives an argument that applies to states indifferently (246a10–246b3). Then he analyzes the cases of states of the body and argues that these are not cases of alteration (246b3–20). He finally addresses the case of states of the soul (246b20–248a9). Here again he has to respect a notorious distinction introduced in his psychology: the distinction of the sensitive and the intellectual part of the soul. He first makes a short argument concerning the states of the soul in general (246b20–247a7). Then he considers the states of the sensitive part of the soul (247a7–19). The last part of the chapter (247b1–248a9) tries to show that neither the states of the intellectual part of the soul nor their acquisition nor their use can count as an alteration. In his closing remark (248a6–9) he comes back to the task he had to achieve according to the opening statement. He affirms that it is evident from his arguments that alteration occurs in perceptible qualities and—this is new—in the sensitive part of the soul and in nothing else.

The Main Problems of VII.3

In this chapter, Aristotle gives reasons for thinking that certain processes which one might naturally have thought were alterations, are actually not alterations after all. Presumably, this is because they seem to involve cases of X altering Y without X being in contact with Y, whereas he had claimed in chapter 2 that every case of change (and therefore alteration) involves contact.

In some ways, this should not come as a surprise. Aristotle's notion of change (κίνησις) is much more restricted than one modern notion of change, namely Russellian (sometimes called "Cambridge") change: if X is F at t1 and G at t2 (where F and G are incompatible), X has changed from being F to being G. Carlo Natali points out that Aristotle's distance from this conception of change is shown by the fact that the processes he denies to be alterations—which fit the Russellian description perfectly—are called by him alterations κατὰ συμβεβηκός (248a9).

So, Aristotle's notion of change is narrower than the Russellian one.

Here are some examples:

(1) If it is true of Socrates at t1 that he is taller than Theaetetus and true of him at t2 that he is shorter than Theaetetus, and this is because Theaetetus has grown, then Socrates has not changed.

(2) If X is at rest at t1 and moving at t2, it is not the case (according to Aristotle) that X has changed from being at rest to moving (because changes take time, and this transition does not take any time).

(3) If X is moving at t1 and at rest at t2, it is not the case (according to Aristotle) that X has changed from moving to being at rest (because changes take time, and this transition does not take any time).

(4) Notoriously, if X is perceiving Y at t1 and not perceiving Y at t2 (or vice versa), it is not the case that X has changed from perceiving to not perceiving (or vice versa).

To change, according to Aristotle, is to undergo a process between limiting states L1 and L2; between being in states L1 and L2, X is changing from L1 to L2. The transition from rest to change is not itself a change, nor is the transition from change to rest. It is not that the state of rest is coming to be, as X slows down. One way of thinking about cases (2), (3), and (4) is that they involve instantaneous switches or transitions, whereas Aristotle is firm in his conviction that changes should take time (as the subject of change changes from being in state L1 to L2).

In our chapter, Aristotle limits what counts as an alteration. In some of the cases discussed in VII.3, Aristotle's denial that they are alterations is related to his denial that the cases listed above are changes. For instance, he says that when somebody comes to know something, this is not a change (i.e., an alteration). When the soul comes to rest after being in turmoil of whatever kind, this coming to rest is not a change. It is the culmination of a change. The body might be churning away (altering)—but the resultant state of the soul (knowing something) does not come to be, any more than the state of rest comes to be.

The body which moves from A to B does just that: it changes from being at A to being at B. Its state of being at rest at B does not come to be. It does not come to be at rest at B—it comes to be at B. Aristotle is clear about this—there is no coming to be of rest (and no coming to be of motion either). Aristotle talks about "acquiring a state." When X arrives at B, X acquires the state of being at rest. It has not undergone a change from being in motion to being at rest, but it has acquired the state of rest where it was previously in motion.

However, the cases discussed at the beginning of VII.3 are not ones whose status of being alterations will be put into question because of their similarity to cases (1) through (4) above. Rather, Aristotle in our chapter makes clear that only some qualities (or some things he treats as qualities in the *Categories*) are such that change in respect of them counts as alteration. In particular, he claims in this chapter that what looks like alteration in respect of shape and state turn out not to be alterations. The first turns out to be substantial change: when a new shape is acquired, in the relevant sense, something new comes into being, so that we do not have an X such that is characterized by one shape at t1, and then another at t2.

Commentaries on the Six Sections of the Chapter

245b3–246a9

Benjamin Morison

Commentary

245b3–9: ***Alteration is more restricted than you think***

Aristotle sets up the problem he wishes to discuss, namely that alteration is only to be found in a restricted number of cases, and in fewer cases than one might have thought.

> That everything which is altered is altered by perceptible qualities, that is, that alteration takes place only in those things which are said to be affected in their own right by perceptible qualities, is to be seen from the following considerations. (245b3–6)

What are "perceptible qualities"? They are referred to just by the term αἰσθητά, but it emerges that they are what the *Categories* calls "affective qualities" (παθητικαὶ ποιότητες). These were (9a28–10a10): sweetness, bitterness, sourness, hotness, coldness, darkness, and paleness. Some of these are called "affective" because they are capable of bringing about states (πάθη) in other things, in particular the senses (9b5–7). Examples of this sort include: hotness, coldness, sweetness, and bitterness. Some are called affective because they have come about through their host having been affected in a certain way (9b9–11). Examples of this sort include: paleness, darkness, and redness.

What is the force of "in their own right"? Its presence is one difference between the second part of the sentence and the first part. The addition of "in its own right" in the second limb of the sentence seems to be adding something to the first: Aristotle is not reformulating the first part, but refining it, to underscore that some things can be affected by perceptible qualities derivatively—but not in such a way as to bring about alteration.

My translation "that is" is not meant to indicate that the second limb repeats the claim of the first limb, but rather refines it. (Gerhard Seel has more to say on this in his contribution; he argues that the change from ἀλλοιοῦσθαι to πάσχειν is also significant).

What would be a case of being affected by perceptible qualities derivatively, but not in such a way as to bring about alteration? Wardy thinks that Aristotle already has in mind the sorts of cases he will go on to describe later, i.e., cases where there is genuine alteration in X (by perceptibles), which results in something bearing a relation to X, namely Y, being affected. So for example, some matter may be altered (condensed or rarefied) and something may thereby come into being:

> It is clear that neither the states, nor the loss or the acquisition of states are alterations, but perhaps it is necessary that they come into being and perish when some [other] things undergo alteration as [it is necessary that] shape and form [is generated when] e.g., warm and cold and wet and dry things [undergo alteration], or those things in which [shape and form] happen to be as in the first things [which they are in]. (246b12–7)

There is another possibility for what Aristotle might have in mind. When he refers to something X being affected in its own right by a perceptible quality, he might be thinking of a case where X is changed by perceptible quality Y, in virtue of Y's altering some part of X. So perhaps I become tanned as a result of sitting in the sun. I have been altered, but not "in my own right"; it is my surface which has been altered in its own right (or perhaps even just a part of my surface).

One oddity in this first sentence is that we are talking about being affected by perceptible qualities, but the second sentence (and the rest of the chapter) is concerned with change in respect of perceptible qualities (which is contrasted with change in respect of shape, etc.). Perhaps Aristotle thinks that if X changes in respect of feature Y, it must have been changed ("in its own right") by a feature of the same sort as Y. See Gerhard Seel's principle PIV, which represents one way of making this thought more precise.

> For of the other [cases], someone would most of all suppose that alteration takes place in shapes and forms, and in states, both in the taking on and casting off of them: but in neither of them [is there alteration]. (245b6–9)

Aristotle claims that changes of shape or form, which you might have thought counted as alterations (245b6), are in fact not alterations (245b8–9). Equally, he says, changes in state are not alterations either. Why does Aristotle take these two cases in particular?

The reason must be this. Alteration is naturally defined as change in respect of quality. But in the *Categories* (chapter 8), shape or outward form is listed as a type of quality (the fourth), as is state (first). Affective qualities—the one true respect in which alteration takes place, according to our chapter—are the third type of quality listed. Clearly Aristotle is restricting the range of qualities which are such that change in respect of them counts as an alteration.

As Simplicius noted, the second type of quality listed in the *Categories*—namely powers, skills, capacities—are not mentioned here in VII.3. He suggests that this is because powers are not qualities strictly speaking anyway. There is no hint of this in the *Categories*. Perhaps Aristotle thought that whatever account one offers of why the acquisition or loss of states does not count as alteration will also explain why the acquisition or loss of capacities does not either.

In any case, when Aristotle says that someone might have been tempted to suppose that changes in respect of form/shape, and condition, are alterations, he must be referring to his own view that these are qualities; the natural identification of alteration with change in respect of quality would then lead you to suppose that change in respect of form/shape or condition would count as alteration.

I have translated the last καὶ of line 245b7 as "both," linking it with the first καὶ of 245b8. Thus, I think that Aristotle is specifying what he means by the curious supposition that states are alterations: what he means to be considering is whether both the taking on of a state, and the casting off of a state, are alterations. Now, one could (of course!) translate καὶ "and," in which case, Aristotle would be imagining someone putting forward the suppositions both that states are alterations, and that the taking on and casting off of the states are alterations. The first of these suppositions is very puzzling indeed, and it seems preferable to suppose that here (and in the passage on excellence in the same chapter, which raises the same question) Aristotle wavers between using the word "state" (and words for states), for the process of acquiring the state (in which case he can just ask whether states are alterations) and for the product of that acquisition (in which case he asks whether the taking on of the state is an alteration). Here I agree with Wardy,[4] but the conference was divided on the issue. There is more on this below, in the commentaries on the passages on excellence.

245b9–246a9: *Change of shape is not alteration*

The general point made in these lines is not hard to discern. Aristotle is keen to show that change of shape is not an alteration. The details of the text, on the other hand, are elusive. The section divides into two, each part giving a consideration in favor of the thesis.

[4] Wardy (1990), 211.

245b9–246a4: *First consideration for change of shape not being alteration*

Aristotle points out that we adopt one locution for the case where something has a form or shape imposed on it, and another locution for the case where something gains an affection (πάθος), i.e., is altered. He takes this difference in locution to be a sign that in the case of the form being imposed, we do not have an alteration.

> For we do not say that that which is being shaped and fashioned, once it has been completed, is that from which it is made. For instance, we do not say that the statue is bronze, or the candle is wax, or the bed wood, but using paronyms, we say that they are brazen, waxen, and wooden. (245b9–12)

"That which is being shaped and fashioned" is ambiguous. When a statue is being fashioned out of bronze and someone asks "what is it that is being fashioned?," you could say with equal propriety, it seems to me, that a statue is being fashioned (from bronze), or that bronze is being fashioned (into a statue). So to which is Aristotle referring here, the bronze (matter) or the statue (end product)? It must be the statue, because the phrase "when it is completed" could not refer to the matter. The bronze is not completed or perfected; the statue is.

The point Aristotle is making, then, is that we do not say that the statue, or candle, or whatever, is "that from which it is," i.e., its matter. We do not say that the statue is bronze, but brazen. We do not say that the final thing (the outcome of the alleged "alteration," the thing which is fashioned, namely a statue) is bronze. We do not say that the final thing (the outcome of the "alteration," i.e., the candle) is wax. The one is brazen; the other is waxen. We do not say that the statue is "bronze with such-and-such a shape." It is not bronze—it is made of bronze. To say that the statue is made of bronze is not to say that it is bronze. (This point is obscured somewhat in English because of the "is" of constitution.)

We say "brazen" instead of "bronze," and Aristotle describes this as a case of using "paronyms." This is not quite a direct parallel to the cases of paronymy familiar from the first chapter of *Categories*. There, two items (e.g., Socrates and virtue) are paronyms, because Socrates has virtue in him and is called "virtuous," where this name is related to the name "virtue" by a change of ending. If X and Y are paronyms then one of X and Y is in the other as a subject (in the technical sense of the *Categories*). However, in our text, there is no such relation. The bronze is not in the statue as a subject; it is the matter of the statue. Aristotle does not use the word παρωνυμιάζοντες in exactly the same sense as in the *Categories*, but in a closely related sense—here, X is called paronymously after Y iff X takes one of its names from the name of Y, but changing the ending of that name of Y (but Y need not be in X as a subject).

Gerhard Seel points out that Aristotle makes a similar point to the one in our passage in a famous passage in *Metaph.* IX.7.1049a18–24. There, Aristotle says that the box (which is made of wood) is not wood, but wooden, and the wood (which is made of earth) is not earth but earthen. Thus, in the *Metaphysics* passage Aristotle is keen to point out that the finished product is not the matter, but does derive its name from it. The point may be similar, but of course the context in the *Metaphysics* is different: Aristotle is seeking to make a point about what is potentially what; he is not concerned there with our question, viz., what counts as an alteration. However, in the *Metaphysics* passage, Aristotle does explicitly draw the parallel with the case where an affection, e.g., whiteness (λευκότης) comes to be in a subject, e.g., a man, and that subject is then named paronymously from the affection, i.e., we call the subject "white" not "whiteness" (1049a29–b2). This latter case is a genuine case of what the *Categories* calls paronymy, so it is understandable that Aristotle uses his word "paronymously" here in *Phys.* VII.3. (It is of course worth pointing out that the *Categories* has nothing to say about matter and form).

Note on the β text: Aristotle says something like "we do not call the form that from which comes the form of the statue." This is an odd way of putting it, since Aristotle must mean that we do not call the composite (the statue, or whatever) by the name of the matter. He seems to be using the form-word to refer to the composite.

> But we do call that which has been affected and altered [that from which it is altered]: for we call the bronze fluid and hot and hard, and the wax too, and not only this way round, but also we call that which is fluid, and that which is hot, bronze, calling the matter homonymously with the affection. (245b12–246a1)

"That which has been affected and altered" is a reference to the thing which underlies the alteration. In the case of an X which is altered from being F to being G, X is the affected thing (or the altered thing). When X is altered from being F to being G, we do say that X (when it is G) is G, and that the G thing is X. In other words, we call the G "that from which it is"—i.e., the unqualified underlying thing, namely X. We are prepared to call X "G," and also the G thing "X." If you call X "G," then "G" is an ὄνομα of X, and if you call the G thing "X," then "X" is an ὄνομα of the G thing.

Put this another way: we do say that the final thing (the thing which is the outcome of the "alteration," namely the hard bronze) is hard. We also say that the final thing (the thing which is the outcome of the "alteration," namely the hard bronze) is bronze.

Ross puts the phrase "and not only this way round, but also that that which is fluid, and that which is hot, is bronze" in parentheses. This must be wrong. Aristotle is trying to make a point here which contrasts with the case he gave in the immediately preceding lines. In those lines, he made the point that when matter X takes on shape Y, we do not call the "outcome" of that acquisition of a shape (i.e., the resulting composite object) "X." The contrasting claim in these lines is that when A takes on affection B, we do call the "outcome" of that acquisition of an affection (i.e., the thing which is relevantly affected) "A." But this latter point is exactly the one made in the "and not only this way round" clause: we call that which is fluid (the thing which has been taken on the affection of fluidity) "bronze." In any case, the way Aristotle expresses himself ("and not only this way round, but also") suggests that the emphasis falls on this claim (that we call the B "A"), and not the first claim (that we call A "B").

Ross was misled, I think, by the final remark "calling the matter homonymously with the affection." He thought this could only refer to the case where (e.g.,) the bronze is called hot, hence he put the intervening claim in brackets. But if A is called homonymously with B, then B is called homonymously with A, and the clause "calling the matter homonymously with the affection" can refer to both preceding claims.

There is a worrying objection to what Aristotle says. We say "the bronze is hot," and "the hot thing is bronze." The underlying thing (the analog of the matter) and the affection share a name. They share a name in that one can use the word θερμόν of both. But why think that the affection is τὸ θερμόν, as opposed to ἡ θερμότης? If the affection were referred to in this way, then there would no longer be "homonymy" between the affection and the underlying thing, but rather "paronymy." But obviously this would ruin Aristotle's point.

A parallel point could be made concerning some of the examples in the *Categories*. When we say that X is white, there is homonymy between the color white and X. This works in Greek if you refer to the color as τὸ λευκόν, but does not work if you refer to the color as ἡ λευκότης. However, this observation does not make much of a difference to the metaphysical concerns of the *Categories*; there, homonymy and paronymy are classed together as revealing cases where one thing is in another as a subject. In VII.3, Aristotle is driving a wedge between the cases of homonymy and paronymy, and arguing that they reveal a crucial physical difference. Hence it does matter in our lines that he takes the affection to be τὸ θερμόν and not ἡ θερμότης.

> The result is that, if in the case of shape and form, that which has come to be is not called the thing in

> which the shape is, but in the case of affections and alterations, [that which has come to be] is called [the thing in which the affection and the alteration is], it is evident that these instances of becoming would not be alterations. (246a1–4)

Aristotle summarizes his points from the previous lines. The translation here more or less mirrors Wardy's and the Oxford translation. Another translation is, I think, possible. Oxford and Wardy take τὸ γεγονός to be the subject of λέγεται and ἐν ᾧ ἐστι τὸ σχῆμα the complement after λέγεται. One could take ἐν ᾧ ἐστι τὸ σχῆμα as complement after γεγονός, to make "that which has come to be the thing in which the shape is"—a complicated but understandable periphrasis referring to the matter—and then take κατὰ τὸ σχῆμα καὶ τὴν μορφήν with λέγεται to make something like the following: "that which has come to be the thing in which the shape is not called according to the figure and form." Gloss: "the matter is not called by the name of the form." If one pursued this interpretation, then the second clause of the protasis would then be translated as follows: "but [that which has come to be the thing in which the affection is] is called according to its affections or alterations."

The salient linguistic difference between this translation and the Wardy/Oxford one is that the κατά clauses are construed each time after λέγεται. The philosophical difference is minimal, but speaks in favor of Wardy/Oxford. On the interpretation being considered here, the first claim is the wrong way round, so to speak—what Aristotle had claimed earlier was that the form (the statue) is not called by the name of the matter (bronze), rather than that the matter (the bronze) is not called by the name of the form (the statue).

246a4–9: *Second consideration for change of shape not being alteration*

> Again, it would seem to be absurd also to say that the man had been altered, or the house, or any of the

> other things which have come to be; but perhaps it is necessary that each comes to be with something being altered, e.g., the matter being condensed or rarefied, or heated or cooled, but those things which come to be are not being altered, nor is the coming to be of them an alteration. (246a4–9)

When a house or a man comes into existence, they have not been altered. Aristotle does not give any reasons for this, but it is not hard to find a good one. The house or the man simply were not around at the beginning of the "alteration" to underlie the change. As Aristotle says, there might have been alteration of something else, but not through having a shape added to it.

It is important to see that this consideration (246a4–9) is complementary to the first (245b9–246a4). According to the first consideration, when matter takes on a new form or shape, the matter should not be said to alter. According to the second consideration, when a new composite object comes into being (through the matter taking on a new form or shape), the object should not be said to alter. Both considerations are considerations in favor of Aristotle's claim that the changes in question are not alterations—they are neither alterations of the matter (consideration 1) nor of the emergent composite substances (consideration 2).

Aristotle clearly allows that in some cases where a composite substance comes to be, there may have been an alteration involved—an alteration, that is, of the matter which underlies the genesis. He does not imply that there is always such an alteration. Gerhard Seel points out that the genitive absolute ἀλλοιουμένου τινός (translated here "with something being altered") may indicate a causal connection: the alleged alteration in question comes about because a different, genuine, alteration comes about in something else.

Questions arising from the passage

In the first consideration, Aristotle appeals to a certain way in which we talk as evidence for a certain metaphysical picture. As noted above in the note on 245b12–246a1, Aristotle has been quite selective in his choice of vocabulary for the relevant affection, and may have made life too easy for himself by taking names such as τὸ θερμόν for the affection (rather than ἡ θερμότης). Moreover, one wonders whether the linguistic point Aristotle makes would carry over to other languages—and what the significance would be if it did not. Presumably, considerations from language are always defeasible—but this is the only argument Aristotle actually offers for a change of shape not being an alteration.

However, Aristotle is surely right that in some cases, when matter takes on a certain shape, what has happened is best described as a coming to be, i.e., as a substantial change. But could it be right that any such taking on of a shape is better described as a substantial change? During the making of a clay statue, there are many intermediate forms or shapes that the clay takes on before it takes on the form of the statue (and the statue comes into being). What are we to make of these intermediate phases of the genesis of the statue? We either have a succession of substances (which is counterintuitive), or we have changes of shape which do not involve a new substance coming into being—in which case we have a succession of counterexamples to Aristotle's claim here.

There are other types of counterexample to Aristotle's thesis which seem easy to come by. The counterexamples I have in mind are changes of shape where one would not be at all tempted to say that a new substance comes into being.

For instance:

(1) I change shape over my existence—I get fatter (regrettably). I seem to change with respect to my shape.

(2) I let the air out of a football—it deflates. It seems to change with respect to its shape.

(3) A flower is buffeted by the wind—it bends. It seems to change with respect to its shape.

Responses:

(1) There seems to be an answer available to Aristotle here: this could be classed as a case of growth. But as David

Charles underlined at the conference, we need to know why it is preferable to class this change as growth rather than change of shape. What is at stake?

(2) There is one observation to be made straightaway, which is that in a certain sense the shape of the football does not change when the air is let out. It is still a spherical ball (suitable for Soccer) or an elliptical one (suitable for American Football); in this sense, the shape would only change if some sort of force were applied to the ball so that its inflated shape would be different (perhaps some sort of kink is introduced). Then, if such a change is brought about, perhaps we no longer have the same ball (and we can say that the original ball has passed out of being). But this does not seem sufficient as a reply to the counterexample, because there is surely another sense in which the deflated ball has changed shape; for instance, we want to explain the fact that the deflated ball does not roll by the fact that its shape is not one appropriate for rolling. To avoid this counterexample, it seems as though Aristotle would have to class this change as a diminution.

(3) Again, in the case of the bent flower, we might make similar remarks; the shape of the flower does not actually change when it bends, unless some damage is done (in which case perhaps the flower does not survive). But again, there is surely another sense in which the flower has changed shape. To deal with this counterexample, we cannot say that a flower has grown (as in case 1), or shrunk (as in case 2). Perhaps the answer then is that the flower has changed place? This is a problematic response, since the flower does not go anywhere. There is a move which can be made here: some changes of place on the part of X, e.g., rotations, do not involve X going somewhere, because they only involve the parts of X changing place—perhaps that is what we have here.

Because of these counterexamples, the mood of the conference was to find Aristotle's arguments insufficient for showing that changes of shape should not be considered to be alterations. Carlo Natali notes that we do not know whether Aristotle would have had more to say about the counterexamples; he is right that the passage is rather condensed, and it would have been nice to have more detail. Gerhard Seel in his contribution offers the solution that predicates of shape are not ordered as contraries along a continuum.

Equally, it might be relevant that Aristotle does not return elsewhere to the bolder claims of this passage (that change of shape is not an alteration, etc.), raising the possibility that he himself did not want to pursue some of these claims further.

246a10–246b3

Ursula Coope

Overview

This section of text provides an argument for the conclusion that states (ἕξεις), whether bodily or psychic, are not alterations. The main argument occurs in the first three lines of our passage. The argument is:

(1) Bodily and psychic states are either excellences or defects. (a11–12)

(2) Neither excellences nor defects are alterations. (a12)

So: bodily and psychic states are not alterations. (a10–11)

Lines 246a12–b3 are a supplementary argument, in support of premise (2).

The argument is:

(a) Excellence is a perfection; defect is a destruction of and departure from this.[5]

Supported by: when something has acquired its own excellence, it is then most perfect.

Supported by:

(i) (When something has acquired its own excellence, then it is most in accordance with its nature).

[5] I discuss what is meant by "this" (τούτου) below, pp. 64–71.

(ii) When something is most in accordance with its nature, then it is perfect. (e.g., when a circle is most a circle and best, then it is perfect).

(b) Neither having nor acquiring perfection is an alteration. (Supported by: it would be strange to call the coping and tiling of a house an alteration, or to say that the house was altered in being coped or tiled).

So: neither acquiring nor having excellence is an alteration (and similarly, neither acquiring nor having a defect is an alteration).

The main questions that arise are:

(1) Why does Aristotle think that all bodily and psychic states are either excellences or defects? (This raises a further question: what, exactly, does Aristotle mean by ἕξις here?)

(2) What does Aristotle think the relation is between the claim that acquiring excellence/defect is not an alteration, and the claim that excellences/defects are not alterations? Does he think that these are two different claims, both supported by the argument he gives, or are these meant to be two different ways of formulating one and the same claim?

(3) Why does Aristotle think that if something is a perfection it cannot be an alteration? (And does his reasoning also show why acquiring a defect cannot be an alteration?) At the conference, we considered three possible answers to this third question: (i) there is no definite beginning for the process of acquiring a perfection, but an alteration must have a definite beginning and end; (ii) the acquisition of a perfection is a process of coming-to-be and hence cannot be an alteration; (iii) when a thing acquires a perfection it merely manifests a nature it already has. I shall defend answer (iii).

Commentary

1. *The first premise of the main argument: states* (ἕξεις) *are excellences or defects*

Aristotle seems to think that this first premise needs no defence. This would not be surprising, if he were claiming simply that excellences and defects are types of ἕξις, but the argument seems to require something stronger than this. The conclusion is that *no* states are alterations, so it looks as if premise (1) needs to be the claim that *all* states are either excellences or defects. But why does he think he can assume this?

We can avoid having to answer this question if we take Aristotle's argument here to have a looser structure than I have suggested. Perhaps all he means to show is that certain central types of ἕξις (namely, excellences and defects) are not alterations. He might think that if he has shown this, he has done enough to make it plausible to conclude that no ἕξις is an alteration, even if he has not given an argument that *proves* this conclusion.

If, on the other hand, Aristotle is assuming here that any ἕξις must be either an excellence or a defect, what might lead him to make such an assumption? We can, perhaps, go some way towards answering this by looking at what he means here by "excellence and defect" and by "state" (ἕξις). He makes it clear that "excellence" and "defect" are to be understood in a broad sense, as including both excellences and defects of the soul and excellences and defects of the body. He goes on to give examples of bodily excellences: health, fitness, beauty, strength (246b4–5, 7–8). He claims (at lines 246a13–15) that when something has its own excellences, it is complete/perfect (τέλειον) and most in accordance with [its] nature (κατὰ φύσιν). His view, then, seems to be that having an excellence is having a standing disposition to function well (whether by having a strong and healthy body or by having good qualities of character). Conversely, a defect will be a standing disposition to function ill.

In the *Categories*, Aristotle describes a state (a ἕξις) as a relatively permanent disposition: "a state (ἕξις) differs from a disposition (διάθεσις) in being more stable and lasting longer" (*Cat.* 8.8b26ff.). If this is the notion of ἕξις that he is appealing to in *Phys.* VII, then the claim that all ἕξεις are excellences or defects is the claim that all of a thing's standing dispositions must either be such as to contribute to, or be such as to detract from, that thing's functioning. To justify this assumption, Aristotle would have to show that there are no ἕξεις that are neutral with respect to a thing's functioning.[6] Perhaps some justification could be provided. For instance, Aristotle could say that, if the possession of a certain disposition made no difference one way or the other to a thing's functioning, then having that disposition would be irrelevant to being a thing of that kind. In that case (he might say), this disposition would be an accidental feature of the thing in question. This might suggest that the disposition could not have the kind of stability that (by definition) a ἕξις has.

However, the fact that Aristotle does not attempt to provide an argument of this sort suggests that in *Phys.* VII he is operating with a different conception of ἕξις. Perhaps he is assuming here that a ἕξις is, by definition, something that contributes, either positively or negatively, to a thing's functioning. We can find some support for this in one of his definitions of ἕξις in *Metaph.* V. According to this, a ἕξις is a "disposition (διάθεσις) in accordance with which the thing that is disposed is either well or badly disposed, either in itself or in relation to something else" (1022b10–12). The example he gives there is health. At least in the *Metaphysics*, then, the notion of being disposed *well* or *ill* is built into the very definition of a ἕξις.

---

[6] Wardy (1990) is, it seems to me, too quick here. He says that, given Aristotle's naturalistic belief that "biological and instrumental kinds alike possess characteristic ἀρεταί shaped by their defining ἔργα," it follows that "any ἕξις whatsoever is up for evaluation, makes a positive or negative contribution to the functioning of its possessor, and thus is virtuous or vicious" (210). The problem with this is that, for all that is said here, a state that is "up for evaluation" might turn out to make neither a positive nor a negative contribution, and hence might be neutral with respect to the functioning of its possessor.

Even if this is right, there remains a puzzle about Aristotle's list of examples at 246b4–5, 7–8. How can Aristotle maintain that (for example) the acquisition or loss of health is not an alteration? Elsewhere, *becoming healthy* is one of his standard examples of an alteration (see, in *Phys.* VII itself, 249a29ff. and 249b5–8, see also *GC* I.4.319b12–13). Is this just inconsistent? It may, perhaps, help here to remember that a ἕξις (and hence an excellence) must be a relatively permanent state. In the *Categories*, Aristotle lists health and sickness as examples of dispositions that are not ἕξεις, but he adds that such dispositions might count as ἕξεις if they became "through length of time ingrained in someone's nature and incurable or very difficult to change," (διὰ χρόνου πλῆθος ἤδη πεφυσιωμένη καὶ ἀνίατος ἢ πάνυ δυσκίνητος οὖσα (9a1ff.). Perhaps, then, Aristotle would say that, when he calls health a bodily excellence, what he has in mind is the relatively permanent state of being in good health; the loss or acquisition of such a state is not an alteration. On the other hand, without losing such a general bodily condition, one might catch or recover from a passing illness. Aristotle need not deny that *this* kind of becoming ill or well is an alteration.[7]

2. *The relation between the claim (a) that acquiring excellence/ defect is not an alteration, and the claim (b) that excellences/ defects are not alterations*

In *Phys.* VII.3, Aristotle seems to be attempting to disarm various apparent counterexamples to the claim that all alteration is in respect of perceptible qualities. Given this aim, it is not surprising that he argues that acquiring excellence/defect is not an alteration. Excellences and defects are not perceptible qualities, so if the acquisition of excellence or defect were an alteration, Aristotle could not maintain that all alteration was in respect of perceptible qualities. This is an important potential objection to his view. After all, he himself, in the *Categories* (8b26–27), says that a ἕξις is a kind

[7] However, Seel (p. 129) suggests reasons why Aristotle might also deny even that the recovery from a temporary illness is an alteration.

of quality (ποιότης), so it would be natural for an Aristotelian to assume that the acquisition of a ἕξις was an alteration.

However, at the start of our section of text, what Aristotle sets out to attack is not the claim (a) that the *acquisition* of excellence or defect is an alteration, but rather the claim (b) that *states of excellence and defect* are themselves alterations. This is puzzling, because it is hard to understand why anyone would suppose that excellence or defect was an alteration (or any other sort of change). On the face of it, states of excellence and defect do not provide even *apparent* counterexamples to the thesis that alteration is in respect of perceptible qualities.

An easy answer to this would be to suppose that Aristotle is simply being careless here: when he says that excellence and defect are not alterations, what he means is just that the *acquisition of* excellence or defect is not an alteration.[8] The word for completion, τελείωσις, perhaps makes it easier to slip between one claim and the other. When Aristotle says that excellence is a completion, τελείωσις, he seems to be saying that excellence is *that because of which* something is complete or perfect. But the word τελείωσις, like the English words "completion" and "perfection" can be used for the acquisition of whatever it is that makes one complete. For example, in the *Meteorologica* (379b18), concoction is called a τελείωσις by heat (and here it must mean a becoming perfect by means of heat). This may help to explain why Aristotle might think that by arguing that a τελείωσις is not an alteration, he has shown that neither having nor acquiring excellence is an alteration.

But there is reason to think that he is not simply expressing himself loosely in this way. His discussion of the house example

---

8 This seems to be Wardy's (1990) view. He suggests that "perhaps Aristotle has been tempted into an inaccurate broadness of expression by his emphasis on the virtues' and vices' character as (im)passivity to change, which is explained in the following sections" (210–211). Morison (above, p. 46), favors a similar interpretation, claiming that Aristotle wavers between two different uses of the word "state" (and also of words for states): (a) for the process of acquiring the state and (b) for the product of that acquisition.

suggests that he is quite deliberately making a claim both about the having and about the acquiring of excellences. He says that it would be odd to think the completion of a house was an alteration. If he were simply comparing *acquiring* excellence with completing a house, one would expect him to say that, just as a house is not altered when it acquires its coping or tiling, so also a person is not altered when he acquires excellence. But, though he does say that a house is not altered when it acquires its coping or tiling, he *also* says that *the coping and tiling themselves* are not alterations (246a18–19).

We are left, then, with the following questions. Why does he think he needs to make *both* claims (and in particular, why is it relevant to emphasize that having excellence or defect is not an alteration)? And what does he think is the relation between these claims?

One possible answer is that the word ἀλλοίωσις (which I have been translating "alteration"), as Aristotle uses it here, is ambiguous. He might be using it both for the process of alteration and also for the state that results from this process. On this view, when Aristotle says that a house's coping is not an alteration, what he means is that the coping is not the result of a process of alteration. (The English word "alteration" can, in fact, be ambiguous in a similar way. "I showed them the alterations to our house" usually means: I showed them the results of the processes of alteration.) If this is right, then Aristotle is not really making two separate points, one about excellence (or defect) and the other about the acquisition of excellence (or defect). He is claiming (i) that acquiring excellence or defect is not a process of alteration, and (ii) (what amounts to the same thing) that neither excellence nor defect is a state that results from a process of alteration.

An alternative answer is that when Aristotle says that excellences and defects are not alterations what he means is that the *exercise* of an excellence or defect is not an alteration. In other words, acting excellently (or defectively) is not being altered. We can see why Aristotle might think it necessary to defend *this* view. It is not, after all, absurd to suppose that someone is altered when he acts excellently or defectively. If this is the right interpretation, then Aristotle is making two different points here: (i) acquiring excellence (or defect) is not being altered and (ii) acting excellently (or defectively) is not being altered.

The choice between these two alternative interpretations depends upon a further question: what is Aristotle's reason for claiming that a perfection is not an alteration? If (as the second interpretation above suggests) Aristotle is making two different points, one about *acquiring* excellence and the other about *acting* excellently, then he must think that the reasons for denying that the acquisition of excellence is an alteration are also reasons for denying that excellent activity is an alteration. I turn now to this further question.

3. *Why is a perfection (or completion)* (τελείωσις) *not an alteration?*

Aristotle argues that since excellence is a completion/perfection and defect is a departure/destruction, neither is an alteration. Before we look at Aristotle's argument for this, we need to ask what exactly he means when he says that defect is a destruction or departure. He says that defect is a destruction "of this" (τούτου). What does τούτου refer to? It is natural to take it to refer to excellence, or to the state of being perfect.[9] But if this is right, then Aristotle is committed here to the implausible view that before acquiring a defect one must have been excellent. An alternative would be to take τούτου to refer to the acquisition of an excellence. The thought then would be that to acquire a defect is to be knocked aside from the acquiring of excellence. But it is quite unclear what Aristotle would say about an occurrence that was the destruction of or departure from an *acquiring*. My preferred interpretation is to take τούτου to refer generally to being in accordance with nature (κατὰ φύσιν). The point, then, would be that becoming excellent is becoming *most* in accordance with nature, while becoming defective is departing from being in accordance

---

9 See Wardy (1990), 211: "Vice is the destruction and departure from this excellence."

with nature. This would allow for the possibility that something could become defective by becoming less in accordance with nature, even if it had not previously been excellent.

Why, then, does Aristotle think that perfections (and departures from what is most in accordance with a thing's nature) cannot be alterations? He supports the claim about perfections by pointing out that it would be absurd to say that for a house to be coped or tiled is for it to be altered. In acquiring coping or tiling, a house is perfected. For a thing to be perfect (or complete) is for it to be "most of all in accordance with its nature" (μάλιστα κατὰ φύσιν). Our question is: why does Aristotle think that these points show that a perfection or departure is not an alteration?

In what follows, I shall look at three alternative answers. According to the first, what distinguishes an alteration from either a perfection or a departure is that an alteration is defined in terms of a beginning and an end (a perfection, in contrast, is defined only in terms of its end). A second possibility is that Aristotle thinks a perfection is a kind of generation, and a departure is a ceasing to be. A third possible answer is that in being perfected or departing from perfection, a thing does not acquire a *new* property, but rather manifests to a greater or lesser degree, a property that it already has.

First answer: a τελείωσις has no definite beginning.

Aristotle's point might be that a perfection, unlike an alteration, is defined solely in terms of its end. Any transition that results in a house's being complete or perfect counts as a perfection of the house. How such a transition starts is not essential to its *being a perfection*. Similarly, what is essential to becoming excellent (a kind of perfection) is that it results in being excellent, not that it starts from any other particular state. If this is the point Aristotle is making, then he is drawing upon the thought that an alteration is a change from some definite state to some other definite state.[10]

One drawback to this interpretation is that it makes it hard to understand Aristotle's remarks about the acquisition of defect. He says that the acquisition of defect is a destruction and departure (246a16–17). But it is hard to read this as denying that the acquisition of defect is a change from some definite state to some other definite state. There is, surely, an end state of becoming defective: namely, being defective. Could Aristotle be denying that there is a definite beginning state? This is a plausible view: provided that someone ends up being defective, he counts as having acquired defect, regardless of what particular state he was in before the acquisition (whether, for instance, he was excellent, or akratic). However, this does not seem to be the point that Aristotle is making here. To characterize the acquisition of defect as a "departure" is, surely, to characterize it as something that does have a definite beginning: a departure is, necessarily, a departure *from* something.

Second answer: a τελείωσις is a coming-to-be, not an alteration.

The example of the house might suggest that Aristotle's point is that being perfected is the last stage of a thing's coming to be. On this view, the reason why being perfected is not an alteration is that the thing that is perfected does not exist until it has been perfected. A change that results in a certain thing's existing cannot be an alteration in that thing, for the thing is not *there to be altered* until the change is over.

In the case of the house, the thought would be that something is not, properly speaking, a house before it has acquired its tiling and coping. But why might anyone think that? One possibility is that without its tiling or coping a building cannot provide shelter. If Aristotle thinks that a house is, by definition, something that provides shelter, then it would follow that a structure without

---

[10] He says elsewhere of change generally, that it is "from this to that," ἐκ τοῦδε εἰς τόδε. (See, for example, *GC* 320a12, on μεταβολή).

coping or tiling would not count as a house.[11] On this view, it is ἄτοπον to say that a house is altered when it acquires its coping and tiling, just because there is not yet a house there to *be* altered, until the coping and tiling have been acquired. Analogously, so the argument would go, someone who lacks the psychic or bodily excellences is not fully a human being. Acquiring these excellences is the final stage in the generation of the human being (just as acquiring coping and tiling is the final stage in the generation of a house). So, it is wrong to say that a man is altered when he acquires the excellences, because he is not, strictly speaking, fully human until he has acquired the excellences.

That this is what Aristotle has in mind is suggested by what he says about shape in the lines immediately prior to our passage (see the commentary by Morison). His point there is that acquiring shape or form is coming to be, and coming to be is not the same as being altered (246a3–4). In support of this, he also invokes the example of a house. He says it would seem odd (ἄτοπον) to say that man, or a house or anything else that has come to be has undergone alteration (246a4–6). Is Aristotle, in our passage, simply making the same point about acquiring an excellence?

This interpretation is also supported by Aristotle's use, in the other version of the text, of the example about the house being coped and tiled. In the β version, Aristotle invokes the example of a house receiving its tiling and coping at 246a28–29. There he explicitly says that what this shows is that "alteration does not occur in things coming to be." If he is making the same point when he invokes this example in the α version, then he must be assuming that in acquiring a ἕξις, a thing is coming to be, and that this is the reason why acquiring a ἕξις is not being altered.

Note that if this interpretation is right, then Aristotle's reasons for denying that acquiring excellence is an alteration are not also reasons for denying that when one acts excellently one

[11] Wardy (1990) claims that θριγκός is the topmost course of bricks, rather than a single coping stone.

is altered. Presumably, he would not want to say that to act excellently is to come into being. This interpretation, then, would force one to reject the second of the two alternatives I suggested in response to problem (2) above: the alternative on which Aristotle is here defending two different claims (both (a) that acquiring excellence is not being altered and (b) that acting excellently is not being altered).[12]

One difficulty with this interpretation (as with the first) is that it is not obvious how Aristotle's argument would apply to the acquisition of a defect. We saw that the coming to be of X cannot be an alteration in X, since X does not exist until the process of coming to be is over. But the acquisition of a defect is (on Aristotle's view) a destruction, not a coming-to-be. Arguably, while X is being destroyed, it *does* still exist: while it is being destroyed, it has not yet been destroyed (and what previously existed and has not yet been destroyed, must surely still exist). If we are to deny that X's destruction is an alteration of X, this must be because X does not persist *after* its destruction. Aristotle's point would have to be that if X is altered, it must be true, when the alteration is over, that X exists in its altered state. But if this was Aristotle's point, it would rely upon the strange view that to acquire a defect was to cease to exist. If Aristotle held (surely more plausibly) that acquiring a defect was the *first part* of a process of destruction (just as acquiring an excellence was the last part of a process of coming to be), then he would have no reason to deny that the same person could persist through the acquisition of a defect.

In any case, the main objection to this interpretation is just that it seems to commit Aristotle to a very implausible view. If acquiring the excellences is becoming a human being, then we only come into being when we acquire the excellences. As Wardy notes (213), this makes it hard to see how anyone could be praised for acquiring an excellence (since prior to acquiring it, he did not

[12] See above pp. 63–64.

exist).[13] Moreover, elsewhere Aristotle himself seems to endorse the common-sense view that to acquire a state is not to come into being. For instance, in *Metaph.* I, he says that Socrates does not come to be when he becomes cultured, μουσικός (983b13–18).

A natural way to answer this on Aristotle's behalf is to say that you persist through the acquisition or loss of excellence, but that it is possible to persist while being more or less fully what you are by nature. When you acquire an excellence, you become more fully the kind of thing that you are (a human being). But if this is Aristotle's view, some new reason is needed for denying that the acquisition of excellence is an alteration. If the man exists (albeit in a less perfect or complete way) before the acquisition of an excellence, then it is not clear why we must deny that acquiring the excellence is being altered. If one and the same thing can persist through a process of being perfected, then simply comparing being perfected to genesis does not give us a reason for denying that being perfected is an alteration. Why is it wrong to say that something is altered when it becomes more fully the kind of thing it is? The interpretation I consider next is an attempt to answer this question.

Third answer: the τελείωσις of a thing is the manifestation of a nature that it already has.

Aristotle could be drawing on the idea that for a thing to be perfected is for its own nature to become more manifest or shine forth. In explaining the claim that excellence is a perfection, he says that when a thing acquires its own excellence it is "most according to [its] nature" (just as a circle is perfect when it is most

---

13 Carlo Natali has suggested to me that this is less of a puzzle if we bear in mind that "excellence," here, should be understood in a very broad sense (as including things like health, not merely states that we are praised for being in). However, even if this is so, Aristotle is presumably committed to the view that there are *some* excellences that we can be praised for acquiring. Also, setting aside this point about praise, it is at least counterintuitive to maintain quite generally that things do not exist until they exist in a perfect state.

circular and best).[14] Perhaps, then, his view is that when a thing is perfected, properties that it *already has* are more fully developed; it does not take on *new* properties. For this reason, perfection is not alteration (ἀλλοίωσις). A thing is altered only if it takes on some property that is different from the properties it already possesses.

On this view, the point about the house (at least, in the α version of the text) is not that a house comes to exist when it receives its coping and tiling. It is rather, that the house, when it is being coped and tiled, is perfecting a nature that it already has. The house, on this view, can be the subject of the acquisition of coping and tiling, but in acquiring its coping and tiling it is not really taking on *new* properties; rather it is perfecting properties it already has.

This interpretation calls to mind *de An.* II.5, where Aristotle suggests that the transition from possessing knowledge to using it is not an alteration. The reason he gives there is that this transition is "a development into the thing's true self or actuality" (εἰς αὑτὸ γὰρ ἡ ἐπίδοσις καὶ εἰς ἐντελέχειαν) (417b6–7). However, it is not clear whether this is quite the same point as he is making in our passage. In *Phys.* VII, Aristotle also wants to maintain that acquiring a defect is not an alteration. Acquiring a defect is certainly not a development into a thing's true self.

This interpretation can make sense of the fact that Aristotle compares being perfected to coming into existence (for instance, in the β version's discussion of the house at 246a28–29). His point is not that when X is being perfected, it does not yet exist. Rather, he is drawing attention to the fact that X's *becoming perfect*, like X's *coming to be*, is a transition towards X's having a fully developed nature. This might suggest that *becoming perfect* was the last part of a process of *coming to be*, but if Aristotle were to accept this, he would have to abandon the natural assumption that when X is coming to be it does not yet exist (since it has not yet come to be).

---

14 As Wardy (1990), 212, points out, the example of the circle is not very helpful, as it is not clear in what sense an imperfect circle is a circle at all.

This interpretation also allows us to explain why Aristotle might think that one and the same argument can show both that acquiring excellence is not an alteration and also that acting excellently is not being altered. Perhaps he would say that I manifest a nature that I already have both when I acquire excellence and also when (having acquired excellence) I act excellently. Neither of these is the taking on of a new property, and hence, neither of them is an alteration.

Can this interpretation explain Aristotle's claim that acquiring a defect is also not an alteration? Clearly, he does not hold that when I acquire a defect, my nature is more fully developed. Perhaps his point is that, in acquiring a defect, I become less fully what I am by nature (without losing my nature altogether). This would allow him to say that just as becoming excellent is manifesting a property that I already have, but to a greater degree, so becoming defective is manifesting a property that I already have, but to a lesser degree.

This third interpretation, then, allows us to answer some of the puzzles we encountered when discussing the two earlier proposals. It does, though, raise difficulties of its own. It is not clear, for instance, what exactly is the difference between acquiring a new property and manifesting to a greater or lesser degree a property that one already has. Suppose, for instance, that something goes from being lukewarm to being hot. On Aristotle's view, this surely counts as an alteration. But why does this count as acquiring a new property, rather than manifesting more fully an existing property (namely, hotness)? (Is it because, in order to become hotter, a thing must be acted upon by something else that is already hot?[15]) Clearly, we need some further explanation of what it is to acquire a new property. In spite of this, this interpretation is, I think, the most plausible of the three I have considered.

Remaining open questions

---

[15] See Seel's principle IV, for an attempt to spell out Aristotle's views on this (pp. 126–127).

I have argued that Aristotle holds that a perfection is not an alteration because he thinks that in acquiring a perfection one merely manifests a nature one already has. If this is right, it raises the following further questions.

(i) Is Aristotle making two distinct claims here: first that the *acquisition* of an excellence is not an alteration, and second that *acting* excellently is not being altered? (As I have said, the argument that a perfection is the manifestation of a nature one already has could be used to justify both these claims.) Or is Aristotle claiming only that the acquisition of excellence is not an alteration (that is, should the remark that excellence is not an alteration be taken to mean that excellence is not the result of a process of alteration)?

(ii) Does it really make sense to think of acquiring a defect as manifesting, to a lesser extent, properties that one already has (rather than as acquiring new properties)?

(iii) More generally, can Aristotle really justify the distinction I have attributed to him here: the distinction between acquiring a new property and manifesting to a greater or lesser extent a nature that one already has?

(iv) Finally, it is natural to think that *Metaph.* IX's discussion of potentiality might help Aristotle to explain what it is for something to manifest more fully a nature it already has. What he seems to need here is the idea that something can be F first in potentiality and then, later, in actual fulfillment. This thought raises two further questions: can the distinctions Aristotle makes in *Metaph.* IX help to solve the problems we have encountered in this part of *Phys.* VII.3? And if so, should we think that when he wrote *Phys.* VII.3, Aristotle had something like this solution in mind, or should we instead suppose that the distinctions of *Metaph.* IX were worked out later, perhaps in response to problems such as those we have encountered in *Phys.* VII.3?

246b3–246b20

István Bodnár

Overview

The passage 246b3–20 can be divided into three shorter sections:

Section 1, 246b3–10, linking excellences (ἀρεταί)—and as will be clear in what follows, also defects (κακίαι)—to relations, and illustrating this with the example of bodily excellences like health, fitness, beauty, and strength, and their opposite defects;

Section 2, 246b10–17, submitting that neither dispositions themselves, nor the loss or the acquisition of dispositions are alterations—dispositions come into being or perish when there is some change in qualities, to which these dispositions are intimately linked;

Section 3, 246b17–20, picking up a point made already at 246b8–10 about bodily excellences and defects, and canvassing the connection between excellences and defects on the one hand, and the alterations an entity undergoes according to its nature on the other.

Commentary

As a preliminary to the discussion of the passage 246b3–20, we should note that the claims the passage makes about excellences and defects, and then more generally about states (ἕξεις, b13) apparently redraw the internal articulation of qualities as presented in *Cat.* 8. That chapter operated with a four-fold distinction, according to which qualities can be (1) states and dispositions (ἕξις and διάθεσις, 8b26ff.), or (2) those qualities which are attributed according to some natural capacity or incapacity (9a14ff.), or (3) qualities of affection/affective qualities (παθητικαὶ ποιότητες, 9a28ff.), and (4) shape and form (σχῆμά τε καὶ ἡ περὶ ἕκαστον ὑπάρχουσα μορφή, 10a11ff.). ἕξεις and διαθέσεις feature both in this passage and in *Cat.* 8, but there is no suggestion in *Phys.* VII.3 that a ἕξις or διάθεσις, like health or illness, could be contrasted to a different kind of quality, being healthy (or: having a propensity for being healthy), and being sickly (or: having a propensity for being ill). *Cat.* 8.9a16–17 explicitly denies that these latter would be dispositions, instead these are asserted "on the basis of having some natural capacity to effect something with ease, or to be exempted from affections" (9a18–9).[16] As the end of sections 1 and 3 make it clear, the excellences and defects of *Phys.* VII.3 make their bearers liable not to be affected, or to be affected in a particular way (246b19–20), i.e., they perform, at least in part, the functions of the second kind of qualities of *Cat.* 8. Accordingly, the distinction of the *Categories*, between ἕξεις and διαθέσεις on the one hand, and powers on the other, could have

---

[16] The conflict between the two treatments of health vs. being healthy is noted by Wardy (1990), 217n86. Wardy, however, does not follow up this line of thought, instead he accepts Simplicius' first suggestion, at 1062, 20–23, that powers are omitted from the list of qualities in *Phys.* VII.3, because it is not clear whether they are qualities. Note, however, that Simplicius' second suggestion, that a further reason might be that the acquisition, and the loss, of these powers is even less perspicuous (conveniently omitted by Ross and Wardy) shows that Simplicius is out of his depth at this point: the fact that the acquisition, and the loss, of powers is even less perspicuous than the acquisition and loss of ἕξεις could just as well have motivated canvassing these powers in *Phys.* VII.3, as they could also be suitable items for the claim that neither they themselves, nor their acquisition and loss are alterations, changes or generations. My proposal should also be contrasted to Ross's second suggestion (674), that the three-fold division of qualities of *Phys.* VII.3 does not include these capacities, because it is earlier than the somewhat more elaborate division of *Cat.* 8.

been dropped in *Phys.* VII.3 by Aristotle as moot, once he realized that ἕξεις and διαθέσεις can perform the function of powers too.[17]

Section 1

One of the key issues in the interpretation of this section, and of the passage as a whole, is to decide whether the introductory claim of 246b3–4, corroborated by the examples of 246b4–6 and b8, states an identity or a relationship of ontological dependence between dispositions and relations. Note that this query does *not* concern how lines b3–4 and b8 should be parsed. As Walter Cavini's intervention stressed at the conference, the expression εἶναι [. . .] ἐν τῷ πρός τί πως ἔχειν (b3–4) and ἐστι τῷ πρός τί πως ἔχειν (b8), should be construed in a way that preserves the parallel with the formulation of *Cat.* 7, that those items are relational "the being of which is identical to being related in some way to something," οἷς τὸ εἶναι ταὐτόν ἐστι τῷ πρός τί πως ἔχειν (8a31–31, see also 8a39–40). The query is rather whether excellences are nothing else but some elements being related in some way to something, e.g., health would be identical to the mixture and due proportion *(analogia)* of warm and cold things, and beauty and strength, and the other excellences and defects would be identical to some similar relation the bearer of the excellence has. Indeed, for some of these excellences, e.g., beauty and strength, this relation could also be the harmonious blend of the constituents, like for health itself.

The alternative to this straightforward identification is that dispositions, excellences, defects are not *identical* to these relations, even though they are ontologically dependent on them, insofar as the dispositions, excellences, defects are, in some sense, constituted by these relations. It is important to stress that the passage in version α leaves this issue open: it starts out by formulating a general thesis, which can be taken either way, and then quotes evidence which can be taken either way as well.

---

[17] See further Morison on 245b6–9, above.

This is already the case in 246b4–6, where "excellences of the body—like health or fitness are placed in the mixture and due proportion of warm and cold things." Here the claim that these excellences are "in" the relationship between hot and cold things submits that these excellences are constituted by these relations. But the strength of what this constitution amounts to remains to be seen. It can be constitution to the extent that there is nothing over and above this relation, that there is an identity between the relation and the excellences and defects; or, alternatively, the claim can be taken to mean that the constitutive relation remains ontologically different from the excellence (or defect) it constitutes.[18]

There can be several considerations for taking these claims about ἕξεις being constituted by relations in the strict sense of identity. One of these could be to parse these claims, and the ones that follow in the α version, along the lines of the straightforward identifications between ἕξεις and relatives propounded in version β. According to version β excellence, and defect, *is in the category of relation* (246a30–31, repeated at 246b24–25), and health *is* some kind of due proportion between hot and cold things, either external or internal (246b21–22). However, this procedure would presuppose that the claims of version α and β amount to the same. Therefore I will rather follow the more circumspect example of Wardy (216, 218), and will retain the ontological "slack" present in the formulations of version α.

Another consideration for accepting the identity of ἕξεις and relatives could be to refer to the claim of *Cat.* 7.6b15–17, where Aristotle submits that excellences and defects, and likewise knowledge and ignorance, are contrary to each other, and accordingly contrariety is to be found between relatives, as excellences and defects *are* contraries. Unfortunately, however, the claim in the *Categories*, with all its straightforward identification, may be of little help, as it does not set out in any detail what sort of

18 Note that this distinction between the relation of the base-line qualities, and the disposition this relation supports is different from the distinction drawn by Viano (pp. 87–88) and de Haas (p. 100) below, between general qualities, which can be relative, and the individual cases which are not relative.

ontological complexity is involved in excellences and defects, on account of which they qualify as relatives. In particular, the fact that excellences and defects are listed together with knowledge and ignorance may suggest that the relation Aristotle has in mind is not some relation among constitutive entities, like the proportion between the hot and the cold, mentioned in *Phys.* VII.3, but rather a relation between the excellence and some further entity which the excellence in question tracks, and its opposite defect fails to track, similar to the relation between knowledge and the object of knowledge, mentioned at 6b32–35. This, however, means that we cannot settle how we should take the claim of *Phys.* VII.3.246b3–4 that "all the excellences are [in the category of] being in some way related to something"[19] by a reference to *Cat.* 7.6b15–17, as it is unclear whether Aristotle has in mind the same kind of relations in the two passages.

Before turning to a discussion of section 2, let me raise another concern: the last lines of the first section supplement the original thesis, that "each of [these excellences] is [in the category of] being in some way related to something" with the further claim that "each of them . . . puts its owner into a good or bad condition in consideration of its proper affections—the ones by which it comes into being or perishes according to its nature" (246b8–10). It is important to stress that this claim need not envisage cases of natural perishing. Instead it may just submit that possession of defects (κακίαι) will predispose the bearer of these defects in ways in which the proper affections the entity has (οἰκεῖα πάθη) contribute to its demise.

It is also interesting to see that instead of preserving, or preservation (σῴζειν—the verb in terms of which version β formulates this claim at 246b24—, or σωτηρία) the final sentence of this section talks about the generation (γίγνεσθαι) of the bearer of excellences, presumably with the supposition that whatever

[19] Here and in what follows I revert to this rebarbative rendering in order to remain neutral on the import of these lines.

contributes to the preservation of an entity also contributes to its generation and *vice versa*.

A further side issue: Jaeger claimed that Aristotle took over the doctrine of bodily excellences and their dependence on proportion from Plato, and that the presence of this doctrine in our passage at 246b4–8 was a sign that *Phys.* VII is an early work. Wardy, rightly, objects that this doctrine can also be found at *Metaph.* XIII.1078a36ff., "hardly a work of the early period" (216n85, on p. 217). But Wardy is not the first to make this observation. He criticizes Ross for accepting Jaeger's suggestion, but this is true only of the first edition of Ross's commentary of the *Physics*, of 1936. In the second edition of 1955, "lithographically reprinted from corrected sheets of the first edition," Ross already rejected Jaeger's genetic argument, quoting passages from the *Nicomachean Ethics* and the *Rhetoric*. It is important to observe that the minor editorial changes on the sheets of the first edition Ross effected were anything but perfect: the *Introduction* still signals, without any change in the second edition, that there are two passages of *Phys.* VII "that have been rightly signalized by Jaeger as marks of an early date" (17), but the comments to both of these passages—to 246b4–8 and to 249b23—are reversed in the second edition.[20]

Section 2

The second section apparently starts out from a very closely-knit connection between relations and dispositions, but this closely-knit connection, once again, need not necessarily amount to identity. The dispositions, excellences, defects this passage speaks about can also be constituted—in one way or another—by some base relation or by a set of base relations. Then the fact that these relations cannot undergo alteration, change or generation in their own right, can be a reason to hold that such dispositions, etc.

---

20 Unfortunately, Wardy (1990), 290, in this instance reads Ross's development the wrong way around, he comments on the note of the first edition about 249b23 as being "in complete contrast to the view expressed in earlier, unrevised editions of [Ross's] commentary."

will also be exempted from those processes of alteration, change and generation in their own right.

First, note that 246b13–14 at the start observes these strictures and Aristotle starts out to formulate his claims by referring to the acquisition (λῆψις) and the loss (ἀποβολή) of dispositions. Even though there is no coming into being or generation (γένεσις) of these dispositions, this does not mean that we would not be allowed to say that in these cases the dispositions "perish" (φθείρεσθαι) and "come into being" (γίγνεσθαι). Rather, the claim is that these dispositions come into being or perish when some other things undergo alteration (246b14–15, see also the similar claim at *de Cael.* I.3.270a27ff.). This claim is linked to the similar claim about the status of shape and form: these also come into being and perish when some other things—like in the case of the coming into being and perishing of dispositions, some hot and cold things, or dry and wet ones, or the ones in which they are directly present (taking over Ross's phrase for the expression ἐν οἷς τυγχάνουσιν εἶναι πρώτοις)—undergo alteration.

I should make two points here. The first is that in this context we should compare the analogous claim about the status of form in *Metaph.* VIII.3.1043b14–18, submitting that "[it] is necessary that this [i.e., substantial form] is either eternal, or it is perishable without actually being in the process of perishing, and it has come to be without actually being in the process of coming to be." *Metaph.* VIII.3 refers to the considerations canvassed in *Metaph.* VII.8, and recapitulated at VII.15.1039b20–27. The comparison of the *Physics* passage and the passages of the *Metaphysics* is interesting for at least two reasons. First, the *Physics* passage not only formulates a similar claim to the one we find in the *Metaphysics*—it also adds an analogous, second claim about dispositions in general. The *Metaphysics* passages—especially the one at VII.15.1039b20–27—refer to the ontological complexity of composite entities, that this ontological complexity is responsible for the fact that composite entities can undergo coming into being and destruction, and in contrast to this neither constituent of this composite entity can undergo similar changes in the same way.

Even though the *Physics* passage stresses the analogy between the two cases, it is clear that the reasons for the claims of our passage cannot be the same: in this example it is base qualities which are compositionally primitive, and they can undergo alterations. The dispositions which are dependent on these qualities cannot be any more primitive than these base qualities themselves. Instead, the reason for the fact that dispositions cannot undergo alteration should presumably be tied to considerations why there is no change of relational entities in their own right, as suggested at *Phys.* V.2.225b11–13. Dispositions are dependent entities, constituted in some sense by the base qualities, and it is these base qualities which have to change in order to make the dispositional entities come into being.

Two further points should be kept in mind when we compare the claims about the lack of generation and perishing, of forms in *Metaph.* VII–VIII, and of dispositions in *Phys.* VII.3. The first is that it is almost inevitable that the "form and shape" (εἶδος and μορφή) Aristotle speaks about in the *Physics* passage is substantial form, and not the fourth kind of qualities mentioned in *Cat.* 8, "figure and shape" (σχῆμα and μορφή, 10a11–12), and so the divergence between the lists of qualities in the *Categories* and in *Phys.* VII.3 does not consist only in the omission of the powers of *Cat.* 8.

All this suggests that no matter whether dispositions retain some of their qualitative aspects—after all, at the end of *Cat.* 8 (11a20–38) Aristotle submits that there may well be entities which are *both* qualities and relations—they clearly form a distinct class from those qualities which constitute their base. Those qualities—or the entities which are qualified by these base-line qualities—can undergo qualitative change, and provide the base-line change as a result of which these dispositions come to be. The hot and the cold, the dry and the wet are expressly mentioned as such.

This again leaves open both options I mentioned at the outset. Dispositions have to be linked to some more basic qualities, but the section does not need to endorse a straightforwardly deflationist ontology. What is ruled out is that such dispositions would be the actual changes of some more basic entities. Instead there has to be some other constitutive factor beyond these qualities—they have to be embedded in a relation. This kind of ontological dependence allows that there may be some further distinction between this relation and the disposition itself.

Two further notes are in order here. The first is that we should be clear about what the alternative Aristotle rejects here consists in. It is important to see that what is canvassed and rejected here is not a full-fledged ontology of flux, but only such an ontology in so far as dispositions are concerned. On this, rejected, view there could in principle be some properties which are not constituted by flux, and they could be the fundamental entities of the world.

The other point is that where I indicate in my paraphrase two options of taking Aristotle's claim at 246b14–17, that either the qualities themselves, or the entities which are qualified by these base-line qualities—can undergo change, the issue is more a matter of what exactly Aristotle's formulations mean, than an issue of philosophical contention. There are no free-floating qualities, any change in a quality is at the same time a change in a qualified entity. Indeed, elsewhere Aristotle speaks unambiguously about the change of a qualified entity: the substantial coming into being is distinguished from, and at the same time linked to the qualitative changes of *matter* ("but perhaps it is necessary that each comes to be with something being altered, e.g., the matter being condensed, or rarefied, or heated, or cooled," *Phys.* VII.3.246a6–8). Note, that the claim that in the case of generation it is matter that changes, i.e., that in this case there is no independently existing substance involved, which would lose some attribute, and receive another one, is parallel to the distinction Aristotle propounds in *GC* I.4.319b6–31.

Section 3

This section forges further links between excellences and defects, on the one hand, and those factors which are responsible for the thing's undergoing alteration, on the other—among them, most notably, nature. Here, the order of causal relevance—as reflected in the way we speak about excellences and defects—is reversed.

It is excellences and defects which make entities susceptible, or not susceptible, of alterations of a certain kind. This is not a completely novel theme in this passage. Some sort of a link has already been introduced between bodily dispositions and those affections (πάθη) which are causally responsible for the coming into being and perishing of the thing (246b8–10), and between shape and form, and alterations of other things (246b15–17). In those two instances the issue of causal priority has not been broached.

Here I would like to turn to the closing sentence of section 3 first. That sentence submits that "each defect or excellence is said in consideration of those [affections] by which that which has [the defect or the excellence] undergoes alterations according to its nature. For excellence makes [that which has it] either liable not to be affected, or to be affected in a particular way, whereas defect [makes it] liable to be affected, or [makes it] liable not to be affected in the opposite way." Here the very beginning of the sentence introduces besides the excellences and defects some further entities, which are causally responsible for changes of an entity according to its nature. Although at this point these entities are invested with causal efficacy, the closing sentence of the section makes it clear that this does not deprive the excellences or defects of causal relevance. The relationship between the dispositions and the entities introduced here—presumably these are those base-line properties which stand in relation to each other and constitute the excellence or defect in question—can be one of *subservience*: the supervening dispositions are entitled to use the causal powers of the entities on which they supervene.

This suggests once again that the passage, in its α version, does not need to propose an elimination of dispositions—excellences and defects—through reduction to relational entities. The links forged between the different types of entities constitute ontological dependence relations, but the fact that supervenient entities can retain their causal priority in contrast to their base means that in this passage of version α of *Phys.* VII.3 they may remain ineliminable, causally efficacious items of an Aristotelian ontology.

Needless to say, this does not carry over to version β. This is further borne out by the fact that the introduction of the base-line entities in the α version is parallel to the claim that version β makes just after the beginning of section 1, that dispositions are relational, because bodily excellences and defects are constituted by the mixture and due proportion of some bodily components (246b4–6). In contrast, version β speaks about changes in some base-line properties (at 247a20–22), but it does not need to contrast the causal efficacy of these base-line properties with the causal efficacy of the dispositions, as excellences and defects are straightforwardly identified with relations in version β.[21]

[21] The final version of this paper was produced with the support of grant K-69217 of the Hungarian Scientific Research Fund.

246b20–247a19[22]

Cristina Viano

Context and division of the passage

As we have seen before, the aim of the entire chapter 3 of book VII is to show that alteration is a change produced only by sensible qualities. Aristotle demonstrates this by excluding the main cases that are wrongly considered to be alterations: (a) the change of the figures and forms; (b) the change of the "states" (ἕξεις) of the body (b.1) and of the soul (b.2). Our passage concerns especially the states of the soul, that is virtues and vices. The conclusion is that acquisition and loss of these states are accompanied by alteration, but they are not alterations themselves.

The passage presents two main sections:

(i) (246b20–247a5): the states (ἕξεις) of the soul are not alterations.

(ii) (247a5–247a19): the coming to be of excellences and defects of the soul is not an alteration but implies the alteration of the sensible part of the soul by sensible things.

First section (246b20–247a5): ἕξεις *are relatives.*

The first section contains arguments concerning two characteristics of the virtues: they are relative and they are perfections.

The states (ἕξεις) of the soul are not alterations because:

[22] I would like to thank Etienne Helmer for kindly translating my text into English.

1) they are relatives: the virtue puts its owner into a good disposition in relation to its own affections, the vice into a bad disposition (247a4–5)
2) virtues are perfections and vices are departures (247a2–3) from perfection.

246b20: ὁμοίως δὲ καὶ ἐπὶ τῶν τῆς ψυχῆς ἕξεων. The link with the preceding section is determined by the adverb ὁμοίως: "The same is the case concerning the states of the soul." But which argument does ὁμοίως exactly refer to? One could think that it refers to the immediately preceding argument, i.e., the argument that the virtues of the body are relatives, because they consist of relations which the underlying elements have to each other or to the environment. These excellences and the acquisition and loss of them are not alterations, but are accompanied by an alteration in the elements (246b3–20).

Actually, ὁμοίως does not refer to this argument only, but to the previous one too (246a10–b3): Aristotle seems to combine two preceding arguments, and to apply them to the excellences (and defects) of the soul.[23] According to the first argument, the perfection of something is not an alteration. According to the second argument something that is in the category of relation cannot be an alteration. Aristotle shows first that the ἕξεις are relatives on the one hand, and excellences and departures from perfection on the other hand (the idea of perfection is conveyed by the word ἀρετή itself = "excellence"), then he argues that they are not alterations, but that alteration is nevertheless present and necessary in the elements and in the sensible part of the soul.

247a1: ἅπασαι γὰρ καὶ αὗται τῷ πρός τι πὼς ἔχειν. Aristotle's thesis is that the ἕξεις are relatives. As a matter of fact, in the *Categories*, Aristotle puts the ἕξεις in the second kind of the relatives (7.6b2 and 8.11a20) on the one hand; and in the first kind of quality

---

[23] I follow Edward Halper's suggestions. I am grateful to him for his comments after our meeting at Vitznau.

(8.8b26–35) on the other. As qualities, the ἕξεις differ from the διαθέσεις by their stability, and they are specifically identified with scientific knowledges (ἐπιστῆμαι) and with virtues (ἀρεταί).

At the end of the chapter on quality (8.11a20), Aristotle remarks—he is probably answering an objection—that the inclusion of the same items in the category of ποιόν and in the category of πρός τι does not lead to an aporia. He shows this by introducing some distinctions. He distinguishes between the use of the corresponding terms as genera (γένη) and their use as singular terms (καθ' ἕκαστα). For instance, if we use the term "science" as genus, we talk about the science of something, for instance the "science of nature." Therefore in this case the term "science" is a relative. But in the case of a singular science such as grammar or the science of music we cannot say "the grammar of x" or "the science of music of x" because that would amount to saying "the grammar of grammar" or "the science of music of the science of music." Therefore in the latter case the term is not in the category of relation, but in the category of quality. If this distinction holds for all the ἕξεις, then we can apply it to virtues as well and make a distinction between virtue as genus and virtue as particular. For instance we can say that this man has the virtue of courage, but we cannot say that he has the courage of something (because the definition of courage already implies that it is the virtue relative to the πάθος of fear).[24]

We should add that generic ἕξεις are relatives of a particular kind. In fact, predicates like big and double (7.6a37–b2) are always predicated in relation to something. We say "Athens is big in comparison to Sparta" or "two is the double of one." This kind of relatives have been called "strong" relatives. However, states are

---

[24] However, we should not overlook the difference between the case of a particular science and the case of a particular virtue. Actually, Aristotle says that we are qualified by a particular science, for we are said to be experts because we possess such or such particular science (8.11a32ff.). However, the same cannot be said of virtue: the virtuous man is not qualified as such by only one virtue: he should possess them all (cf. *NE* VI.13 on the unity of virtues).

relatives of a different kind, they are relative to something (like knowledge is relative to something) but they are also something else, for instance, concerning knowledge, they are also qualities. Therefore these relatives have been called "soft" relatives.[25] From the perspective of the *Categories*, the generic ἕξεις are then considered as "soft" relatives, that is relatives whose being does not consist only in its relation to something else, because they are also qualities.[26]

These distinctions between genera and species, "strong" and "soft" relatives, are specific to the *Categories*, and their interest goes further than the topic and purpose of our passage, but they are helpful to understand in what sense virtues and vices, which are qualities that qualify us, can be considered as relatives too. From this point of view, one can see that even the perfections and departures from perfection are here species of relatives, since the τελείωσις and (mostly) the ἔκστασις are always of something.[27]

It is important that Aristotle says explicitly what the entities are to which virtues and vices are relative: they put their owner into a good or bad disposition (διατίθησι) with regard to his "own affections."

247a3: πρὸς τὰ οἰκεῖα πάθη. How must we understand the expression: "its own affections"? Does the term "affections" refer only to physical affections, or do they refer to psychic passions, such as anger as well? The first hypothesis is supported by Aristotle's statement in 246b10: "by 'proper' affections I mean those that naturally promote or destroy the existence of something" (cf. Simplicius, *in Phys.*, 1067, 23: "The 'proper affections' are the ones 'by which,' when duly proportioned, [bodies] come into being and

---

[25] On the distinction between "strong" and "soft" relatives: Sedley (2002), 324–352, namely 334.

[26] On the anaysis of the ἕξεις in the *Cat.*, I am grateful to Walter Cavini for his comments after our meeting at Vitznau.

[27] Walter Cavini drew my attention to the fact that the question of the extension of the relatives to all the non-substantial realities still remains unanswered. See Ackrill (1963), *ad Cat.* 7.6a36, 98–99 and more recently Bodéüs (2001), 119.

are in good condition, [namely] hotness, coldness, dryness, wetness, and such, but, when disproportioned, bodies perish and become corrupted, becoming more easily affected and easily changed in their proper affections by [their] surroundings and by one another"). The second hypothesis is supported by the fact that our passage is concerned with virtues and vices, of the soul (cf. Simplicius again, ibid., 1071, 18: "It is obvious that the virtues of the soul are also relatives, since they too are instances of due proportion of those things in which they arise—emotions, and appetites [θυμῶν καὶ ἐπιθυμιῶν] and such—just as the bodily [virtues are instances of due proportion] of hot, and cold [bodily constituents] and such").

Actually, as the rest of the text, which looks so evidently "reductionist" and materialist, will show, the proper affections of virtue and vice are, ultimately, physical. They amount to physical pleasures and pains, which are alterations of the sensitive organs.

Second section (247a5–247a19): *"an excessively physiological account" of the genesis of virtue and vice.*

The second part of the text is certainly, as Wardy suggests,[28] the most interesting and problematic. Aristotle makes a distinction between the generation of virtue and vice on the one hand, and alteration on the other hand. But he states at the same time that this generation implies the alteration of the sensitive organs by sensible objects. Aristotle gives more strength to this statement as he underlines the fact that every ethical virtue is concerned with bodily pleasures and pains, which all flow ultimately from sensation (247a7: ἅπασα γὰρ ἡ ἠθικὴ ἀρετὴ περὶ ἡδονὰς καὶ λύπας τὰς σωματικάς).

According to Ross, ad. loc., this statement means that all the pleasures are ultimately linked to the body. But if this is the case, how can the expression ἅπασα ἠθικὴ ἀρετὴ be consistent with the distinction between the bodily pleasures and the psychic pleasures mentioned in *NE* III.13.1117b28?

---

[28] Wardy (1990), 221.

Actually, in *NE* III.13, analyzing σωφροσύνη, Aristotle distinguishes bodily pleasures (σωματικαί) from psychical pleasures (ψυχικαί). The latter, for instance, love of honors or erudition occur without bodily alteration. In fact, the ethical virtues and vices that are mentioned here (temperance and intemperance) regard only the physical pleasures, and among them, the more primary and "tactile," which are peculiar to animals too and concern touch and taste (cf. 1118a26). The pleasures of sight and hearing are excluded. As we will see later, the "physical" reduction of pleasures in *Phys.* VII.3, is not inconsistent with *NE* III.13. On the contrary, it is complementary to it because it offers a physical explanation to some aspects of Aristotelian ethics.

This second section can be divided into two parts, (a) (247a6–14), and (b) (247a15–19). Wardy[29] considers (a) as the set of premises from which Aristotle elaborates his argument in (b):

(a) (247a6–14)
- 1) Every ethical virtue is concerned with bodily pleasures and pains (a7–8).
- 2) These pleasures and pains are to be found either in action or in memory or in expectation (a8–9).
  - 2.1) As a matter of fact, those which are to be found in action depend upon perception and then the subject is altered according to sensible quality (a10–11);
  - 2.2) and those which are to be found in memory and expectation flow from sensation as well (a11–12).
    - 2.2.1) For we feel pleasure in remembering an affection we once felt or in expecting an affection we may feel in the future (a12–13).
- 3) All pleasures of this kind flow from the influence of perceptive qualities (a13–14).

[29] Ibid. 222ff.

4) When this occurs the sensible part of the soul is altered by the sensible things (a7).
5) Then virtues are necessarily generated when a change occurs in the sensitive part the soul (a6–7).

(b) (247a15–19)[30]

Since 1) virtue and vice concern pleasure and pain (247a15–16);
2) virtue and vice arise when pleasure and pain do (247a14–15).

But 3) pleasures and pains are alterations of the sense organs (247a16–17).

Therefore 4) it is necessary that virtue and vice are lost and gained when something (i.e; some sense organ or part of some sense organ?) undergoes alteration (247a18–19).

According to Wardy (223ff.), the argument (b) raises two main problems about the presentation of the conception of pleasure in this passage:

(i) the statement that pleasure and pain are alterations of sensitive organs;
(ii) the form and strength of the dependence of virtue and vice on pleasure and pain.

Indeed, the whole passage gives a materialist or, rather, a sensualist conception of pleasure. Ross (676) already noted that the argument according to which the sensible part of the soul is altered by sensible qualities leads Aristotle to give "an excessively physiological account of the genesis of virtues and vices." Virtue and vice look as if their sole function were to form or decrease the sensitivity to bodily pains and pleasures. Ross refers to the *Nicomachean Ethics* where Aristotle changes this perspective when he says that the attitude of men towards pleasure and pain can be altered by the love for one's friends or homeland, or

[30] For the sketch of (b), I follow Wardy (1990)'s division, 222ff.

by the desire to achieve the moral good (τὸ καλόν). As Ross also points out, at *NE* 1174a13–175a3, pleasure and pain are not defined as alterations but as realizations and loss after the alteration of a sensitive organ.

In order to give an account of this problem, Simplicius (1073, 11–22) suggests that where Aristotle says that pleasures and pains are alterations, he actually means that pleasures and pains *supervene* to alterations involved in perception. As examples of alterations of the αἰσθητικόν, he refers to cooling and heating, expansion and compression, and he says that the affections go with pleasures and pains. When they are well balanced, then virtues follow; when they are not, then vices follow.

Wardy, 224, explains the reduction of pleasure to alteration by the fact that Aristotle must keep here a delicate balance: on the one hand, he must deny the fact that the change in the ethical ἕξις is an alteration; on the other hand, he also wants to convince us that there is a close connection to alteration. The aim of his argumentation requires that we consider only the changes in the "affective qualities." A more precise elaboration of the specific function of these alterations in the perceptive process would be unclear in this context. That is why Wardy suggests to see this identification of pleasure and pain with a psychic alteration not as a conception of pleasure independent of *NE*[31] but as "an excusable simplification of a more subtle idea."

Accordingly, the idea of pleasure one can find in *Phys.* VII.3 can be taken as a simplification and a restriction of the more comprehensive analysis of *NE* (X.3 for instance). The reason for such a simplification has its justification in the context of *Phys.* VII.3: Aristotle wants to show that alteration by the sensible qualities plays an important role in the generation of psychic states and what exactly this role is; but he does not want to identify this alteration with these states. Therefore we can understand why he prefers the σωματικαί pleasures and does not mention the ψυχικαί

---

[31] For example, Gosling and Taylor (1982), 194–195, consider this text as evidence for an early Aristotelian/Platonic view of pleasure.

pleasures, that occur without bodily alteration (*NE* III.13.1117b28). In that sense, the phrase ἅπασα ἠθικὴ ἀρετή at 274a7 could refer to σωφροσύνη, as a paradigmatic example of virtue, whose specific object is precisely the bodily pleasures, and more precisely those of the sense of touch (sex, drink and food).

On the alteration and the affective qualities

Actually, in *Phys.* VII.3, the notion of alteration too is defined in a narrow way: it occurs only through the "affective qualities" (παθητικαὶ ποιότητες). In *Phys.* VII.2.244b5, the παθητικαὶ ποιότητες are said to be responsible for the alteration in a body and they are presented as the properties according to which a being can be subject to affections:

> For our assumption is that things that are undergoing alteration are altered in virtue of their being affected in respect of their so-called affective qualities; for every body differs from another in possessing a greater or lesser number of sensible characteristics or in possessing the same sensible characteristics in a greater or lesser degree. But the alteration of that which undergoes alteration is also caused by the above-mentioned characteristics, which are affections of some underlying quality. Thus we say that a thing is altered by becoming hot or sweet or thick or dry or white; and we make these assertions alike of what is inanimate and of what is animate, and further, where animate things are in question, we make them both of the parts that have no power of sense-perception and of the senses themselves.[32]

The commentators (cf. Simplicius, and Ross, 673) identified these affective qualities to the third kind of qualities as described in

[32] The passage is corrupted, I follow the text established by Ross on the basis of Simplicius' commentary.

the *Categories* (8.9a28): the παθητικαὶ ποιότητες, called as such because they qualify the objects that received them.

As Wardy (163) pointed out, the reduction of alteration to affective qualities in *Phys.* VII.3, even if Aristotle does not present it as a deliberate reform of the classification one can find in *Categ.* 8, has serious consequences. As a matter of fact, the limited idea of alteration in *Phys.* VII.3 can easily be seen as the best fundamental explanatory principle of the science of nature as such. According to Wardy (162–163), if Aristotle has limited the alteration to affective qualities, it is because he considered these qualities as logically simpler than the other kinds of qualities listed in *Categ.* 8 (states, conditions, natural capacities, and incapacities, forms). On the one hand, these qualities can affect the senses really and directly: sensations are not subjective *epiphenomena*, they correspond to qualities that exist in the outside world. On the other hand, they are simple because, as their analysis shows, they do not contain any sort of complexity—complexity which is proper to the states or conditions of which they can be the constitutive parts.

I do not agree entirely with this remark about "simplicity"of the affective properties. As a matter of fact, in the *Categories*, Aristotle makes a distinction between two kinds of affective qualities: the first are called "affective" not because the thing that possesses them has been affected when it first got them, but because these qualities are able to affect our senses (for instance honey, that produces a sensation of sweetness), the second are called "affective" because they are also produced by some passion of the soul (such as blush due to shame) (9b33). Now the first are simple while the second are not. The latter can qualify an individual temporarily or permanently, and they determine his way of undergoing the corresponding affections (cf. the proper affections, τὰ οἰκεῖα πάθη, of the soul and the body as well, *Phys.* VII.3.247a3 and 246b10):

> Similarly with regard to the soul also we speak of affective qualities and affections. Those which are present

> right from birth as a result of certain affections are called qualities, for example, madness and irascibility and the like; for in virtue of these people are said to be qualified, being called mad and irascible. Similarly with any aberrations that are not natural but result from some other circumstances, and are hard to get rid of or even completely unchangeable; such things, too, are qualities; for in virtue of them people are said to be qualified. (translation by J. L. Ackrill)

In particular, those that qualify the individual permanently originate at birth from some affections of matter (ἐν τῇ γενέσει εὐθὺς ἀπό τινων παθῶν, 8, 9a35). Aristotle refers to madness (lit.: "loss of reason by madness": μανικὴ ἔκστασις [τοῦ λόγου]) and anger as examples. Simplicius (*in Categ.* 259, 21) insists on the fact that these affective properties are constitutive, and are part of the *substratum*: they are πάθη that altered matter at birth and they left marks, they became constitutive in the sense that they make the individual able to "catch" the πάθη that occur. For instance, the angry man, who at birth underwent an alteration in the matter of his body (an alteration of his blood that made it thick), undergoes the affection of anger in a peculiar way that is more intensively than anyone else. The affective qualities of this kind reveal a certain level of complexity because they are at the same time the origin of an affection and the result of an undergone affection.

I believe that this notion of affective qualities that matches "the proper affections" is the key for understanding when and how alteration plays its parts in the generation of the ἕξεις of the body and the soul. This idea of "passive receptivity" is the basis of Aristotle's hylomorphic theory of passions and character, and it is fundamental not only for his psychology but also for his physics that is his theory of animated and unanimated beings of the sublunar world.

In none of Aristotle's works is the theory of "affective qualities" systematically developed: affective qualities are mostly approached through "crossing" analysis, in different contexts, in such a way that at first sight we get a rather complex or even inconsistent and amazing view of this notion. But as any reader familiar enough with the texts of Aristotle knows, the best method to investigate this notion is by taking into account the specific characters and problems of each text. But this is the subject of another paper.[33]

Note on the *textus alter* (β)

> [246b27] Nor is there any alteration involved with psychic virtues and vices, since virtues is a sort of perfection (each thing is most perfect and according to nature when it hits upon in proper virtue, just as a circle [b30] is most according to nature when it is most circular). Vice is the destruction and departure for these <perfections>. And although the acquisition of virtue and the loss of vice occur when something undergoes alteration, nevertheless neither of these is alteration: but that something is altered is clear. For virtue is either a type of impassivity or passivity in a given manner, while vice is either passivity or a susceptibility to affection opposite to that engendered by virtue. And all ethical virtue resides in pleasures and pains, [a25] since pleasure is either in actualisation or through memory or from anticipation. Thus if it is an actualisation, then perception is the cause, while if it is in actualisation, then perception is the cause, while if it is through either memory or anticipation, it is derived from perception. For the pleasure occurs either in our recollection of how we have been affected, or in our anticipation of how we shall be affected.
> (246b27–247a28, translation by Wardy, 59)

[33] In particular: "États du corps, états de l'âme. Aristote et le mécanisme commun des qualités affectives," forthcoming, in Barnes and Bonelli.

On the whole, version β does not say anything different from the α version, but it looks more concise. However, the organization of the developments is different:

(1) The identification of virtues to relatives is not exposed here, but above (246b21).

(2) We find the thesis that virtues are perfections. At 246b28–30, the sentence between brackets that gives an account of the perfection according to nature and refers to the circle as an example, can be found above in α (245a13–15).

(3) The passage 247a22–23 on passibility and impassibility in virtue and vice echoes a passage above in α, 246b17–20:

> περὶ ταῦτα γὰρ ἑκάστη λέγεται κακία καὶ ἀρετή, ὑφ' ὧν ἀλλοιοῦσθαι πέφυκε τὸ ἔχον· ἡ μὲν γὰρ ἀρετὴ ποιεῖ ἢ ἀπαθὲς ἢ ὡδὶ παθητικόν, ἡ δὲ κακία παθητικὸν ἢ ἐναντίως ἀπαθές.
>
> For each defect or excellence is said in consideration of those things by which that which has it undergoes alterations according to its nature. For excellence makes that which has it either liable not to be affected, or to be affected in a particular way, whereas defect makes it liable not to be affected in the opposite way.

But as Ross points out (731), the second version β (ἢ ἐναντία πάθησις τῇ ἀρετῇ) differs from α, which reads: ἡ δὲ κακία παθητικὸν ἢ ἐναντίως ἀπαθές (246b20) ("or impassive with the impassivity which is contrary to the passivity bestowed by excellence," Ross, 675). This second version seems clearer than the corresponding passage in α, in which the place of ἐναντίως is problematic.

Conclusion

The most interesting points of this passage are the following:

(1) In section 1: the inclusion of the ἕξεις in the category of relatives. Due to a comparison with the *Categories* we

can identify them with "soft" relatives, in particular the ἕξεις are relative to the "proper affections."

(2) In section 2: the "materialist" and "reductionist" identification of pleasure and pain with the alteration of the sensitive organs. Actually, this physiological account is considered by some commentators as excessive or oversimplified. But it gives prominence to the hylomorphic view of psychology and ethics and shows the continuity uninterrupted from the body to the genesis of virtues and vices in the soul.

(3) Concerning the text: the version from the *textus alter* (β) offers a more concise reading, and, in some case, a more clear reading of the same passage.

247b1–247b13

Frans A. J. de Haas

Commentary

In this section of *Phys.* VII.3 Aristotle continues his argument designed to show that no alleged modification of the soul is an alteration strictly speaking. He now turns his attention to the states (ἕξεις) of the intellectual part of the soul. He claims both [1] that these states themselves are not alterations, and [2] that there is no generation proper of such states (247b1–7). For the first time in *Phys.* VII.3, Aristotle adds that [3] there is no generation of the use and actuality of the state either (247b7–9). Here he employs a distinction he makes in *de An.* II.5, viz., between the presence of a state, e.g., the possession of a piece of knowledge (first actuality), and the use of the state in the application of that piece of knowledge in the act of thinking about a particular case (more details below). The final paragraph focuses on the very first acquisition of knowledge rather than the state that results from it, and emphasises [4] that the original grasping of knowledge is itself neither a generation nor an alteration (247b10–14).

The first claim [1] echoes the introduction of the topic at 246a10–11. The claim is at first sight puzzling because Aristotle tends to regard states (ἕξεις) as relatively stable qualitative conditions which are not themselves a process of any kind. The *Categories* chapter on quality distinguishes ἕξις from διάθεσις in this way, with instances of understanding (ἐπιστῆμαι in the plural) and excellences as prime examples of such ἕξεις; see *Cat.* 8.8b27–35, 9a4–10. *De An.* II.5.417b15–16 recalls this distinction, to which we shall return below. In so far as knowledge is considered a quality (ποιότης), it may come up for discussion under the heading of qualitative change or alteration (ἀλλοίωσις). However, in the *Categories* Aristotle specifies that only particular instances of knowledge are to be ranked under quality, because qualitative descriptions like "knowing" or "knowledgeable" (ἐπιστήμων) rest upon a person's having a particular instance of knowledge. On the generic level, however, such states and dispositions belong to the category of relatives, just like excellences and defects (246b3–4, 247a1–2). Like sense perception, knowledge is relative to its object (*Cat.* 8.11a20–38). Hence knowledge (ἐπιστήμη in the singular) is duly listed as relative in respect of the object known in *Cat.* 7.6b2–6; 33–35; 7b22–35.

In *de An.* II.5.416b33–35 Aristotle claims for perception that it occurs "in being moved and undergoing" (ἐν τῷ κινεῖσθαί τε καὶ πάσχειν) and is therefore considered to be a kind of alteration (δοκεῖ ἀλλοίωσίς τις εἶναι). On the basis of the parallel between perception and thought which characterizes Aristotle's method in *De anima,* he suggests that thinking (τὸ νοεῖν), too, might consist in a kind of πάσχειν (*de An.* III.4.429a13–15). The chapter continues by specifying the difference between πάσχειν in perception and thought respectively, with a most striking claim that thinking occurs entirely apart from the body (*de An.* III.4.429b4–5). This would already rule out thinking from being associated with any ordinary kind of alteration. However, Aristotle chose to be notoriously unclear about the special status of thinking in this regard. Throughout *de An.* II–III Aristotle labors to specify the relevance for perception and thought of changes in the material sense organs.[34] In this wider context it makes sense to deny that a state of the intellectual part of the soul is itself an alteration. On even more general terms, the exclusion of motion from the states of the soul in *Phys.* VII.3 contributes to Aristotle's aim from *de An.* I.3 onwards of showing why his predecessors were wrong in assuming that the soul moves the body by being in motion itself.

---

[34] This is a vexed issue in Aristotelian scholarship. For a cautious summary of the debate, with references to *Phys.* VII.3, see Magee (2000), 121–149.

On the contrary, so Aristotle argues, motions start from or end in the soul. Our passage in *Phys.* VII.3 can be read as providing the more general physical underpinnings of the view of the soul expressed in *de An.* I.4.408b13–18:

> It is doubtless better to avoid saying that the soul pities or learns or thinks, and rather to say that it is the man who does this with his soul. What we mean is not that the movement is in the soul, but that sometimes it terminates in the soul and sometimes starts from it, sensation e.g., coming from without, and reminiscence starting from the soul and terminating with the movements or states of rest in the sense organs.[35]

According to the second claim [2], in 247b3–7, there is no generation of intellectual states of the soul. The β version has "And this is clear" (τοῦτο δὲ δῆλον) instead of "And further" (ἔτι δέ), thus suggesting a closer connection between the first and second paragraph. This is quite right insofar as the argument constitutes an example of the claim that knowledge is relative. The generation of a state of the intellectual part of the soul is an example of the realization of a *first* actuality from the lack of that state, or first potentiality. *Second* actuality (the actual thinking of a particular object) is not referred to as ἕξις in Aristotle. Nevertheless, in *de An.* II.5.417a31ff., Aristotle doubted whether it would be at all appropriate to speak of "alteration" (ἀλλοίωσις) in both cases that will be under discussion here, i.e., the transition from potentially knowing to actually knowing as in knowledge acquisition (first potentiality to first actuality), as well as the transition from the possession of knowledge to the actual use and activity of knowledge in its application to a particular case (second potentiality to second actuality).

The argument in 247b4–7 regards the generation of knowledge in a soul which turns a potential knower into an actual knower.

---

[35] All translations of Aristotle are taken from J. Barnes, ed., *Revised Oxford Translation* (Princeton, 1984), with slight modifications.

Aristotle states that this transition is not a change properly speaking, because it is caused by the presence of something else, viz., a particular object. The β version reads "for from experience of the particular we take knowledge of the universal" (ἐκ γὰρ τῆς κατὰ μέρος ἐμπειρίας τὴν καθόλου λαμβάνομεν ἐπιστήμην), thus providing a more obvious link to sense experience as the source of universals.[36] The presence of a particular somehow causes the potential knower to know the universal that resides in the particular. For, according to *de An.* III.8.432a3–6 intelligible forms reside in perceptible forms. In other words: when a particular has come to be present (whatever changes that process might entail) the required relation between the potential knower and its proper object, the universal in the particular, comes to be.[37]

Since Aristotle has already argued in *Phys.* V.2.225b11–13 (invoked at 246b10–12, 247a1–2) that there is no proper change in the category of relatives, the reduction of the generation of an intellectual state to the presence of its co-relative immediately excludes alteration (as well the other three modes of change). It is puzzling, however, that this argument is applied to the alleged coming to be of what must be *particular* states of the soul, which, in the *Categories* at least, *did* count as qualities rather than relatives (see above).

In *de An.* II.5.417b19–28 Aristotle explains that the objects of perception are particulars, which are external to the soul; cf. *Phys.* I.5.189a5–8. Hence perception, as well as knowledge of perceptibles, depend on the presence of their object, which they do not have at their disposal. Knowledge of intelligibles is different in two ways: it concerns universals, and universals are somehow present in the soul; hence the mind can focus its attention on any universal whenever it wants. A grammarian who is actively thinking about a particular alpha, is concerned primarily with

---

36 For discussion of the different readings see Wardy (1990), 227ff.

37 It is a moot point whether Aristotle considers the form in a particular as a particular, a universal, or both. For a concise discussion of the problem see Politis (2004), 251–254; for more details see e.g., Woods (1991), 41–56, and Frede-Patzig (1988).

this particular alpha, even if he does so by knowing it qua alpha, i.e., qua the universal alpha which is in the particular alpha that instantiates it. In this sense, knowledge, too, consists in a relation with an object that has to be present. As Aristotle states in *Metaph.* XIII.10.1087a15–21:

> For knowledge, like knowing, is spoken of in two ways: as potential and as actual. The potentiality, being, as matter, universal and indefinite, deals with the universal and indefinite; but the actuality, being definite, deals with a definite object; being a "this," it deals with a "this." But indirectly sight sees universal color, because this individual color which it sees is color; and this individual alpha which the grammarian considers (θεωρεῖ) is an alpha.

From this text it seems clear that knowledge depends for its functioning on the presence of universals that inhere in sensible forms (see *de An.* III.8.432a3–6). The mind manipulates representations of these in the imagination which Aristotle considers indispensable for human thought (see *de An.* III.7.431a14–15; b2–10; III.9.432a7–12). Ultimately, however, all knowledge derives from sensible particulars that are present to the individual, who—as 247b4–7 explain—is acquiring knowledge through sense perception. In *APo.* I.18 Aristotle states in the clearest possible way that universals derive from sense perception and induction, and that therefore all knowledge, whether acquired by induction or by proof, depends on sense perception. It is to this necessary condition of knowledge that our passage draws attention.

In the third paragraph (247b7–9) Aristotle proceeds to the claim [3] that the transition from states of the intellectual part of the soul to their use and actuality is no generation either. As pointed out in the introduction, it is for the first time in *Phys.* VII.3 that Aristotle addresses the generation of second actuality. The reader is invited to consider such transitions as being similar to opening one's eyes (ἀνάβλεψις), and touching (ἀφή). What is the point of comparison? Wardy has argued that in all of these cases there is no *process* of generation; *a fortiori* their use will involve no real change.[38] It is hard to see why the lack of a *process* of generation would entail that the use of the result involves no real change either, especially since the use of knowledge occurs *after* it has come to be. Nor does it seem possible to distinguish between "generation" and "use" in the cases of opening one's eyes and touching at all.

At this point, one of the rare parallels for ἀναβλέπειν and cognates in Aristotle is illuminating. In *History of Animals* VI.3.562a17–21 Aristotle describes the various stages of the growth of a bird within the egg as they can be observed when breaking the shell at different intervals. He states:

> During the period above referred to the chick sleeps, but if it is moved it wakes, looks up and chirps (ἐγείρεται δὲ καὶ ἀναβλέπει κινούμενος καὶ φθέγγεται); and the heart and the navel together palpitate as though the creature were respiring.

This text suggests that waking up and opening the eyes, or looking up, are closely connected to a motion (κινούμενος) with which they are not identical. Waking up is Aristotle's favorite example of second actuality (see *de An.* II.5 *passim*; and below *Phys.* VII.3.248a6). For touching a similar story can be told: two things have to have moved towards each other before they actually touch, but the touching itself is not a motion with a beginning and an end. Things either touch or they do not. The point of comparison, then, is rather that looking up and touching are examples of actualities which are the result of antecedent motions but are themselves without coming to be—precisely because no process is involved in their actualization.

The analog of touching has a further application in Aristotle's philosophy since it is sometimes used with reference

[38] Wardy (1990), 234 (his emphasis).

to the thinking of non-composite substances or essences. In his exploration of being in terms of truth and falsity Aristotle states that thinking or asserting are similar to contact, ignorance to the lack of contact. Non-composite substances exist actually, and do not come to be. One cannot be in error about them; one can only think them or fail to do so. The passage is worth quoting in full:

> Truth or falsity is as follows: contact and assertion are truth (assertion not being the same as affirmation), and ignorance is non-contact. For it is not possible to be in error regarding the question what a thing is, save in an accidental sense; and the same holds good regarding non-composite substances (for it is not possible to be in error about them). And they all exist actually, not potentially; for otherwise they would come to be and cease to be; but, as it is, being itself does not come to be nor cease to be; for if it did it would have to come out of something. About the things, then, which are essences and exist in actuality, it is not possible to be in error, but only to think them or not to think them. (*Metaph.* IX.10.1051b23–32).

Since thinking is contact between the mind and its object (see e.g., *Metaph.* XII.7.1072b18–21), *qua* contact the state itself does not come to be nor cease to be, even though reaching this state may depend on antecedent motions of various kinds.

The final paragraph of this section (247b9–13) returns to the original grasp of knowledge, which one would be most inclined to regard as some kind of generation or alteration. Aristotle has recourse to the familiar claim that to know and to have understanding obtain when our reasoning faculty (διάνοια) has come to rest. This is crucial, for if this is what knowledge and understanding amount to, Aristotle can directly apply the results he argued for in *Phys.* V.2 to the effect that there is no process of generation that leads to rest, nor generation of any change whatsoever.

> Nor is there motion in respect of agent and patient—in fact there can never be motion of mover and moved, because there cannot be motion of motion or generation of generation or in general change of change.
> (*Phys.* V.2.225b13–16)

These claims are argued for at length in *Phys.* V.2. How is change of change to be conceived?[39] A change cannot serve as a substrate of change, nor as a form that is replaced with another form in the same change (225b16–33). If change is itself changed, and generation generated at the moment of their occurrence, an infinite regress arises that points to the lack of a first change; if there is no first change, all change will be abolished (225b33–226a6). In general, the same thing is capable of contrary motions as well as rest, and the same thing is capable of both generation and rest. So something which is in the process of coming to be coming-to-be (as in "generation of generation"), will be perishing once it *has* come to be coming-to-be, i.e., its opposite, or rest (226a6–10). Moreover, it is impossible to conceive of a substrate for the change of change, or the generation of generation. If the generation of learning (μάθησις) cannot itself be a case of learning, nor will the generation of generation itself be a case of generation (226a10–16). Finally, which of the three kinds of change apart from substantial change, viz., alteration, growth and diminution, or locomotion, will be at stake if change changes (226a16–23)? These are the arguments Aristotle relies on in support of his claim that the first grasp of knowledge is neither generation nor alteration.

How are knowledge and understanding a state of rest? More than once Aristotle hints at a connection, also on the etymological level, between knowledge (ἐπιστήμη) and coming to a standstill (ἐφιστάναι), or more generally between the reasoning faculty (διάνοια, τὸ ἐπιστημονικόν) and resting (ἠρεμεῖν): cf. *de An.* I.3.407a33–34; III.11.434a16–21; *APo* II.19.100a6–b5; *Int.* 16b19–21;

[39] For a full discussion of the problems involved see Sorabji (1979), 159–177.

and below, *Phys.* VII.3.247b17–248a6.[40] The most famous account of knowledge acquisition along the lines indicated here is found in *APo.* II.19:

> So from perception there comes memory, as we call it, and from memory (when it occurs often in connection with the same thing), experience; for memories that are many in number form a single experience. And from experience, or from the whole universal that has come to rest in the soul (the one apart from the many, whatever is one and the same in all those things), there comes a principle of skill and of understanding—of skill if it deals with how things come about, of understanding if it deals with what is the case. [. . .]
>
> What we have just said but not said clearly, let us say again: when one of the undifferentiated things makes a stand, there is a primitive universal in the mind (for though one perceives the particular, perception is of the universal—e.g., of man but not of Callias the man); again a stand is made in these, until what has no parts and is universal stands—e.g., such and such an animal stands, until animal does, and in this a stand is made in the same way. Thus it is clear that it is necessary for us to become familiar with the primitives by induction; for perception too instills the universal in this way. (*APo* II.19.100a3–9; 100a14–b5).

As a whole, this text may serve as a summary of the various themes invoked in *Phys.* VII.3.247b1–13, even though its emphasis is different. *APo* II.19 aims at showing how the discriminatory power of perception[41] is fundamental to the acquisition of universal concepts and renders innate knowledge superfluous (against Plato).

---

[40] For a useful account of our passage from this angle see Tracy (1969), 274–275.

[41] For the importance of the discriminatory powers of the soul see de Haas (2005), 321–344.

Our section of *Phys.* VII.3, on the other hand, exploits the fact that this process *comes to rest* in universals: intellectual states of the soul are states of rest which are thereby themselves excluded from alteration and generation properly speaking.

247b13–248a9

Carlo Natali

Division of the passage

Our passage can be divided in three parts. In the first one, ἔτι . . . ἕξιν, 247b13–17, Aristotle proposes the comparison between acquiring knowledge for the first time and recovering knowledge from a temporary loss of it, due to sleep, drunkenness or similar states.

In the second section, τῷ γὰρ . . . καὶ ἐγερθῇ, 247b17–248a6, Aristotle explains why he thinks that the two cases are similar. This is the apparently weaker part of the argument, and has been much discussed.

In the third section, φανερὸν οὖν . . . συμβεβηκός, 248a6–9, Aristotle draws a general conclusion that summarizes the results of the chapter, concentrating on the last section, from 246b20 onwards, but not excluding what precedes.

The argument

Aristotle here presents a further argument (ἔτι) for the thesis that the state of knowledge in the intellectual part of the soul is not the fruit of a generation: οὐδ' ἔστιν αὐτῶν γένεσις (247b2). In the β version there is no ἔτι and also the reference to book V is absent. The connection of thinking is less clear, and the first argument is missing.

I put into parenthesis the section 247b18–248a2, i.e. "(that's why also children are incapable of either learning or discriminating by means of the senses as well as their elders, i.e., because the disturbance and change [during childhood] are great)," in order to have a clearer connection between the general principle established at 247b17–18, i.e. "it is by the soul's coming to rest from its natural disturbance that someone comes to be wise and knowledgeable" and the distinction put forward at 248a2–6, i.e. "and it settles down and comes to rest, with regard to some objects because of nature itself, with regard to others under other influences." The passage I put into parenthesis articulates in detail what is said in general at 247b17–18. Some interpreters, however, connect the distinction at 248a2–6 only to 247b17–248a2 (see Philoponus, 771, 10, and Carteron, 83), but the limitation is not necessary.

In the preceding section, as Simplicius aptly notes (1079, 30–34), Aristotle describes the passage from potential to actual knowledge as a sort of a "Cambridge change." In fact the knower does not change, but something, the knowable object, enters in a relationship to him. After a comparison between actualization of knowledge and actualization of sensible perception, like touching, Aristotle assumes the case of acquiring knowledge under a general principle, according to which there is no generation of generation or of rest, a principle already established in book V (225b15). The argument is:

(i) since the original acquisition of knowledge is a coming to rest, and
(ii) "coming to rest" is not a generation, then
(iii) the original acquisition of knowledge is not a generation.

Here the preceding passage ends.

In our passage Aristotle presents an argument in favor of (i) by means of a double comparison. The first one is between a man who acquires knowledge for the first time and a man who recovers knowledge from a state of sleep, illness or drunkenness. He says that in both cases there is no alteration in the soul, but in both cases there is a bodily alteration that does not involve the soul. We have here a case of "Cambridge change." The soul does not change, but in the body occurs a modification, i.e., some sort of obstacle is removed. This allows the soul to function (cf. Simplicius, 1078, 15–16, and 1079, 1–5, Wardy, 153–154, Manuwald, 97–98). A second comparison clarifies what happens. It is the same as when someone has their eyes wide open in absolute darkness and cannot see anything: it suffices that some light arrives, and he immediately sees without changing. Again a case of "Cambridge change."

Aristotle concludes that the change is in the body (248a4), and only *per accidens* in the intellectual soul (a8–9; on this point see also Verbeke, 254–255).

The first section, 247b13–17

> Furthermore, (a) just as when someone has passed from drunkenness or sleep or illness to the opposite condition, we do not assert that he has reacquired knowledge by a process of generation (despite his having previously been incapable of using his knowledge), (b) so neither [does this occur] when someone acquires the state in the first place.

The Greek here is ambiguous: Aristotle makes a comparison (ὥσπερ, 14) between what we do not assert in the case of the drunken man, and what we do not assert in the case of the first acquisition of knowledge. But what do not we assert in both cases? On one interpretation, followed by the ancient commentators and some moderns, in both cases we do not say that the man *acquires* knowledge because he already has it. On the other interpretation in both cases what we do not say is that the man acquires knowledge *by a process of generation.*

The first interpretation is implied in the problem discussed by Simplicius at 1079, 6–7: εἰ λαμβάνει ποτὲ τὴν ἕξιν τῆς ἐπιστήμης καὶ φρονήσεως μὴ πρότερον ἔχοντα, πῶς οὐκ ἐπιστήμονα γίνεται ἐξ ἀνεπιστημόνων. Some of the modern translations seem to follow the first interpretation: "we do not say that one becomes possessed of knowledge" (Hardie and Gaye); "neither . . . we do say that he becomes a knower" (Wardy); "we do not say that one becomes a knower" (Hagen), "de même ne nous disons pas non plus . . . <qu'il est devenu savant>" (Pellegrin).

The second interpretation can be found in Ross, 425, and some translations are very clear in pointing out that what is excluded here is the idea of "generation": "de meme il n'y a pas non plus de génération" (Carteron), "neppure possiamo parlare di generazione" (Russo), "asi tampoco puede hablarse de generacion" (Boeri). Ruggiu and Manuwald are more general: "non è così neppure all'inizio" (Ruggiu); "so sagen wir das auch nicht" (Manuwald), but Manuwald immediately goes on to explain, 96, that what is excluded here is the process of generation and not the fact that someone changes from knowing to not knowing. I think that the second interpretation must be right.

The second section, 247b17–248a6

> For (c) it is by the soul's coming to rest from its natural disturbance that someone comes to be wise and knowledgeable (d) that is why also children are incapable of either learning or discriminating by means of the senses as well as their elders, i.e., because the disturbance and change [during childhood] are great) (e) and it settles down and comes to rest, with regard to some objects because of nature itself, with regard to others under other influences. In either case (f) the processes are accompanied by the alteration of certain things in the body, just as in the transition to use and activity, when someone becomes sober and wakes up.

As already said, I put into parenthesis the section (d) in order to have a clearer connection between the general principle established in (c), and the distinction put forward in (e) (cf. also Manuwald, 97, and Pellegrin, 371). The (e) passage articulates better what is said in general in (c). Some interpreters, however, connect the distinction only to what immediately precedes, see Philoponus, 771, 10, and Carteron, 83, but the limitation is not necessary. Ross, 677, seems to refer the distinction in (f) both to the general principle (c) and to the children's example of (d).

Here we have the reason why we do not make the assertion mentioned above. The acquisition of knowledge is a case of coming to rest, and we already said that coming to rest is not a generation. It is a case of coming to rest because the soul stops being disturbed by the body. When the blocking physical phenomena are removed, the person acquires or re-acquires his knowledge, because his soul is not disturbed any more.

There is a bodily change, an alteration, that stops the soul's disturbance (cf. Philoponus, 877, 23–24: γίνεται γὰρ ἡ τοιαύτη δύναμις ἐν τῇ ψυχῇ οὐκ αὐτῆς παθούσης τι οὐδὲ μεταβαλούσης, ἀλλὰ τῆς φυσικῆς ταραχῆς καταστάσης). In the case of the person who ceases to be asleep or drunk (ὅταν νήφων γένηται καὶ ἐγερθῇ) we pass from simple possession to actualization of knowledge (ἐπὶ τῆς χρήσεως καὶ τῆς ἐνεργείας). Some manuscripts have ἐγέρσεως, "in the case of waking up", here, and Bekker and Prantl prefer this reading, but Ross reads χρήσεως, with the best manuscripts, and so do I.

Aristotle distinguishes two kinds of causes of coming to rest: human nature and external influences. In the β version we have a temporal reference (τοτὲ μὲν . . . τοτὲ δὲ, 248a27–28) instead of the distinction of two causes, but the reference is temporal only in appearance. Aristotle has in mind the same cases (on the difference of the two sections see also Wardy, 235).

Simplicius (1080, 5) identifies the first case with perception and the second case with desires and practical knowledge. Modern scholars have accepted in general his interpretation (cf. Ross, 677, and Pellegrin, 371), and so do I, but not limiting the first case only to perception. In fact, disturbance has to do also with memory, and is an obstacle also to what derives from memory, ἐμπειρία and τέχνη or ἐπιστήμη. In *Mem.* 1.450a32–b2, Aristotle says that "memory does not occur in those who are subject to a lot of movement, because of some trouble or because of their time of life" (καὶ τοῖς μὲν ἐν κινήσει πολλῇ διὰ πάθος ἢ δι' ἡλικίαν οὖσιν οὐ γίγνεται μνήμη, transl. Sorabji) and universal knowledge is based on memory, as Aristotle says both in *Metaph.* I.1 and in *APo.* II.19.

The second case is when a strong desire perverts wisdom, as Aristotle says in *EE* 1246b14–15. To overcome it we need a good character and a good upbringing as Aristotle says in *NE* 1179b20–21. There is a false problem in this passage. The best illustration of it can be found in Wardy's book, 236, where he says: "The central premise of this argument is the claim that we can seriously draw a parallel between original acquisition of an epistemic ἕξις and recovery from some debilitating condition which temporarily occludes the mental faculties, otherwise the comparison is a mere metaphor and the reasoning collapses." On this point Wardy has forerunners and followers, see Simplicius, 1079, 1–5; Philoponus, 877, 19–28; Boeri, 154–156.

This interpretation is connected with the understanding of the preceding passage, 247b13–17. If Aristotle's point is to deny that in both cases, drunkenness and first acquisition of knowledge, the person involved becomes a knower at all, the argument is irremediably weak. In fact it is simply not true that recovering a forgotten truth and acquiring it for the first time are the same thing. Between a drunken jet pilot and me there is a difference, even if here and now we are both unable to drive an Airbus 310. Aristotle describes very well the difference between the two states in *de An.* II.5 and in *NE* VII.5. There is a difference between (1) having the capacity of acquiring knowledge, (2) having acquired knowledge and not using it, and (3) having knowledge and using it.

It is true that there are differences between *Phys.* VII.3 and *de An.* II.5, in particular the fact that Aristotle in *de An.* II.5.417a31 admits that acquiring knowledge is a form of ἀλλοίωσις, and here he denies it. But all the attempts to explain how it is possible that when someone first acquires knowledge he does not become a knower seem not convincing to me. Simplicius, 1079, 6–14 refers to Plato's theory of recollection, or to Plotinus' idea of "having the intelligible but not having it ready to hand." Both positions cannot be attributed to Aristotle. Philoponus, 877, 19–22, seems to deny that there is acquisition of knowledge because the knowledge is potentially already in the knower. It is not completely clear to me that this is Philoponus' position: if we consider the

Greek fragments of his commentary, we are allowed some doubt. The Arabic paraphrase of Philoponus' commentary is closer to it but adds a distinction between children and young men that is not in the text, cf. 770, 10–771, 3. However, this is how Boeri, 155, interprets Philoponus, and he may be right. In any case, this position denies Aristotle's distinction between state (1) and (2) as well. Wardy's position attributes to the man who first acquires knowledge a disposition or readiness to learn, and his exegesis seems not very different from Philoponus'.

A simpler way of getting out of trouble seems to be the following. On the one hand we admit that Aristotle in *Phys.* VII.3 already knows the difference between not having, and having and not using, because it would be very strange if he did not. On the other hand, we should admit that there is a difference between the positions in *Phys.* VII.3 and *de An.* II.5 concerning the understanding of the concept of ἀλλοίωσις. In *Phys.* VII.3 Aristotle uses a very narrow concept of ἀλλοίωσις, as many interpreters say (cf. Manuwald, 49–50, with further references): that is why he denies here that coming to know is an ἀλλοίωσις, while he admits it in the *de Anima.* So, we are not forced to attribute to Aristotle the strange theory that the drunken jet pilot and I are in the same situation with regard to driving an Airbus 310. The similarity of the two cases is limited only to one aspect, i.e., the fact that in both cases *there is no generation* of knowledge. But they differ in many other senses, not relevant here.

This can be accepted, if we consider both cases under the aspect of the relation between knowledge and the body, and we do it in a very general way. In both cases the knowledge is impeded by some bodily movements, and when these movements stop the knower can either acquire or re-acquire his knowledge. I say "in a very general way" because the bodily movements in the young age are similar to drunkenness but not quite identical to them (cf. *NE* 1154b9–11).

The example of the παῖδες, in (d) illustrates both the idea that to know one must eliminate the natural disturbance in the body and the idea that this rule applies to the first knowledge as well as to recovering it when it is lost for a moment.

The third section, 248a6–9

> Thus (g) it is evident from what has being said that being altered and alteration occur in perceptible qualities and in the sensitive part of the soul, and in nothing else except *per accidens.*

Here Aristotle refers principally to the section about the sensitive parts of the soul, but also to the other sections of the chapter. It repeats what he has said at the beginning. Compare the two passages:

> ἐν μόνοις ὑπάρχει τούτοις ἀλλοίωσις ὅσα καθ' αὑτὰ λέγεται πάσχειν ὑπὸ τῶν αἰσθητῶν, ἐκ τῶνδε θεωρητέον. (245b4–6)

and

> φανερὸν οὖν ἐκ τῶν εἰρημένων ὅτι τὸ ἀλλοιοῦσθαι καὶ ἡ ἀλλοίωσις ἔν . . . τοῖς αἰσθητοῖς γίγνεται . . . ἐν ἄλλῳ δ' οὐδενί. (248a6–8)

There is a clear correspondence between the two passages that open and close the chapter. But there is also something new. First of all, having explained the connection between what happens in the material parts of the physical objects and the living bodies, on the one hand, and the appearance of shapes, forms and states, on the other, Aristotle can now add the clause κατὰ συμβεβηκός. In fact he has demonstrated that the functioning of the mind is not an alteration, but it is connected to an alteration of the body. That alteration is a *conditio sine qua non* of the functioning of the mind.

Besides, in the opening clause Aristotle defines alteration in terms of what causes the alteration without specifying what entities undergo alteration. On the contrary, in the closing clause he specifies the entities that undergo alteration: perceptible qualities and the entities that have these qualities on the one hand, and the sensitive parts of the soul on the other. This latter point is very important, because here we have a difference between the two parts of the soul, one undergoing alteration, the other not (I owe this second observation to Gerhard Seel).

Mind-body problem in Aristotle?

If we judge from the present passage, Aristotle's account of the relationship between mind and body in the process of acquiring knowledge and using it is the opposite of modern tendencies. Today neuroscientists often prefer to describe thinking and deciding as physical processes of the brain, and define the conscious level of our thinking as "supervenience," i.e., as an epiphenomenon that has no causal impact on the functioning of our brain. Aristotle on the contrary in *Phys.* VII.3 seems to think that mind and soul are active *per se,* and the influence of the body can only harm or facilitate this functioning, but is not causally responsible for it. But a wider analysis of Aristotle's texts would be needed to fully appreciate this difference and to show the superiority of his psychology to the modern errors and mistakes.

Appendices

The Logical and Semantic Background of Aristotle's Argumentation

Gerhard Seel

The purpose of this article is to explain how Aristotle's argument in *Phys.* VII.3 runs from a logical point of view and what its semantic preconditions are. I shall first make a conjecture concerning Aristotle's definition of alteration, then I shall try to confirm this conjecture by relevant passages from further texts inside and outside the *Physics.* Finally I will analyze Aristotle's argument in *Phys.* VII.3 in light of these findings.

I. *A conjecture concerning Aristotle's definition of alteration*

Using formal language the two principles that Aristotle sets out to prove can be given as follows:

PI For all x, if x is altered, there is a y in such a way that y is a perceptible quality and x is altered by y.

$$(x)\ (A'x \rightarrow Ey\ [Qy \cdot xAy])$$ [42]

PII There is no x such that x is altered and there is no y in such a way that y is a perceptible quality and x is affected by y in its own right.

$$-Ex\ (A'x \cdot -Ey\ [\ Qy \cdot xPy\ ])$$

[42] A'x means "x is altered," A' being a one place predicate; xAy means "x is altered by y," A being a two-place predicate; xPy means "x is affected in its own right by y." Qy means "y is a perceptible quality."

PII differs from PI in two points: 1. It has a negative existential quantifier instead of a positive universal quantifier and the dependent formula has an adjunction instead of an implication and a negative universal quantifier in front of the second variable instead of a positive existential quantifier. This would mean that the two principles are equivalent unless there were differences in the meaning of the predicates, but a closer inspection shows that there are such differences. 2. In fact there are two differences of this kind: a) the predicate "x is altered by y" is replaced by the predicate "x is affected by y"; b) the qualification "in its own right" is added to this predicate.

For these reasons PII is not simply a more precise reformulation of PI. What is the exact relation of the two principles? PII explains how alteration is brought about, namely by affection. Thus PII gives the reason for PI. Does this amount to a definition of the term "x is altered by y" in the following manner?

PIII For all x and all y, x is altered by y iff x is affected by y in its own right.

$$(x)\,(y)\,(xAy \leftrightarrow xPy)$$

No doubt, PIII does not logically follow from PI and PII. However, the question is whether Aristotle would have subscribed to such a definition. The following counterexamples make us doubt this: If a snowman is affected by heat he melts away. If a piece of bronze is affected by enough powerful heat it loses its shape and becomes liquid. In both cases something is affected by a perceptible quality, but neither is a case of alteration according to 245b9–246a4. Can we rescue the hypothesis that Aristotle meant a definition in the sense of PIII by interpreting the expression "in its own right" in a restricted way that excludes the above-mentioned cases?

II. *Evidence from further texts*

At the beginning of book V of the *Physics* Aristotle takes much care to explain the difference between "accidental (κατὰ συμβεβηκός) change," "change by change of a part of it" (τὸ δὲ τῷ τούτου τί), and "change in its own right" (τῷ αὐτό). The latter is also called change καθ' αὑτό (224b22). The example for the first is "something musical walks"; for the walking thing is musical by accident; the example of the second is "the body is healed because his eye or chest is healed"; for the third Aristotle does not give an example, but he underlines that a thing that undergoes this kind of change must in itself be capable of being changed in this specific way. It must have this specific capability according to the different kinds of change and in each kind according to the specific type of change. So in the case of alteration the thing must in itself be capable of being altered in a specific way, for instance, capable to be healed or being heated (224a28–30). Aristotle emphasizes that the same distinction holds for the thing that causes the change (224a30–34). Because change presupposes something that is changed, something that causes the change and two properties, one from which and the other into which the change takes place (224a30–b1), the distinction applies to the latter as well (224b16–22). Aristotle makes it clear that all three elements of the changing process must be individuals and not universals. Therefore he says that when something that is becoming white becomes also an object of thought, the later change is accidental; but when we say that it changes in color we have to do with the second type of change; for white is a part of the genus "color"; but only the change from an individual color to another individual color—say black to white—is a change in itself.

Let us now apply these distinctions of kinds of change to the case of alteration. Though we have no passage in the *CA* where Aristotle formulates explicitly a definition in the sense of PIII, we get crucial information concerning this from some passages that, taken together, allow us to confirm our conjecture.

The first is *GC* I.4.319b6–14 where Aristotle defines the difference between generation and alteration:

> Concerning coming to be and alteration we have to say in which point they differ, for we say that theses changes are different from one another. Now because one thing is the substratum and another is the affection whose nature

> it is to be predicated of the substratum and because there is a change of both of them, alteration occurs, when the substratum, persisting as a perceptible entity, changes in its affections which are either opposed to one another or one intermediate between two extremes.

Here we get the following important clarifications: 1. The subject of alteration, that is the thing that is altered, changes in its affective qualities, but persists as such, i.e., in its essential perceptible qualities, during the process of alteration. 2. The quality the substratum has at the beginning of the process of alteration (the quality it loses) and the quality it has at the end of the process (the quality it acquires) are placed on the same scale or range of qualities as extreme or intermediate qualities. (See also 319b32–320a2).

Aristotle gives the example of a body that—though persisting as the same body—is now healthy and now ill and the example of a piece of bronze that is now spherical and at another time angular (b12–14). The second example does not seem to fit the second point very well, for the properties "spherical" and "angular" are not contraries lying on the same scale of properties. But what is more troublesome with these examples is the fact that they do not correspond to the more restrictive conception of alteration we found in *Phys.* VII.3, which explicitly denies that change of state and change of form are to be counted as alterations.

Aristotle clarifies his concept of alteration further in *Phys.* V.2.226a26–29:

> Let change concerning quality be alteration. I take the term "quality" not in the sense of the quality that is a definitional part of essence (for difference is a kind of quality), but in the sense of the affective quality in regard of which we say that something is affected or resists affection.

I think that what is affected or resists affection is the underlying substratum and this affection causes the change of the affective qualities of the substratum and this change is the alteration.

We get a further important clarification in a passage of *Phys.* VII.2, where Aristotle argues that the thing that undergoes alteration and the thing that causes alteration do not have anything intermediate between them:

> Nor again is there anything intermediate between that which undergoes and that which causes alteration. That becomes clear by induction. For in every case we find that the ultimate point of that which causes alteration and the first point of that which undergoes alteration are in the same place. For our assumption is that the things that undergo alteration are altered by being affected in respect of the so-called affective qualities; for every body differs from some other body in respect of his perceptible qualities either possessing a greater or lesser number of different qualities or possessing the same qualities in a greater or lesser degree. However, the thing that undergoes alteration is also altered by the above-mentioned [affective qualities]. For these are affections of the underlying quality.[43] Thus we say that that which becomes hot or sweet or thick or white undergoes an alteration. (244b2–8)

The picture we get here is the following: A thing is altered in one of its perceptible qualities if it comes into direct contact with something else that possesses a different perceptible quality of the same scale or the same quality in a greater or lesser degree in such a way that the perceptible quality of the latter is the cause of the change in the former. In a passage of *De generatione et corruptione* quoted below, Aristotle speaks of assimilation. Examples would be water becoming hot by being in direct contact with fire, vinegar becoming sweet by being mixed with honey, water becoming white by being mixed with white color, a thin liquid becoming thick by being mixed with a thick liquid. That

[43] The expression "affections of the underlying quality" in b6 is puzzling. Wagner (1983), 651, gives the example of color and red, red being an affection in the range of color. I follow his conjecture.

mixture implies alteration of both things that are mixed each affecting the perceptible qualities of the other can be seen from *GC* I.10.328a32–34 and especially from the final remark of the chapter (b23–24): "mixture is unification of things susceptible to be mixed resulting from their alteration."

This picture is confirmed by the passage *GC* I.7.323b29–324a9 where Aristotle argues that two things must belong to the same kind and differ in species in order to affect one another.

> But since only those things which either involve a contrariety or are contraries—and not any things selected at random—are such as to suffer action and to act, agent and patient must be like (i.e., identical) in kind and yet unlike (i.e., contrary) in species. (For by nature body is affected by body, flavor by flavor, color by color, and so in general what belongs to any kind by a member of the same kind—the reason being that contraries are in every case within a single identical kind, and it is contraries which reciprocally act and suffer action).

He adds a little bit further down in the text (324a10–12):

> For this reason it is well explained why fire heats and the cold thing cools, and in general why the active thing assimilates to itself the patient.

On the basis of these clarifications it seems appropriate to make the following conjecture concerning the meaning of the expression "being affected in its own right":

PIV For all x and all y, x is affected by y in its own right iff

1. y is a perceptible quality;
2. x undergoes a change in at least one of its perceptible qualities u;
3. this change is caused by a direct contact with something that possesses the perceptible quality y;

4. the perceptible qualities u and y belong to the same scale of qualities, i.e., to the same kind of quality;
5. the quality that x acquires by this causal process is either identical with y or is placed between u and y on a common scale.

If this conjecture were correct[44] the following cases would not be cases of affections by a perceptible quality in its own right and thus not cases of alteration either:

1. Iron becomes red by being heated in fire.
2. Wine turns into vinegar.
3. Milk becomes sour.

The first process is not an affection in its own right because the qualities hot and red do not belong to the same kind of quality. In the two other cases the process is not caused by direct contact with something that already possesses the relevant quality.

These considerations make it plausible that Aristotle would in fact have accepted PIII as a definition of alteration and that in this definition the term "being affected in its own right" must be understood in the sense of PIV. It also is, as we shall see, the only interpretation that allows excluding change in shape and change in state from the field of alteration.

III. *The arguments of chapter 3*

Let us now see how Aristotle executes his announced strategy to prove the two principles by showing that apparent counterexamples do not in fact refute them. One would expect Aristotle to start off with the totality of possible cases of alteration. But

[44] We must take into account, however, that our interpretation of the expression "in its own right" is not the only one that is possible. Wardy argues—with reference to *Metaph.* V.18.1022a14–17 and 1022a29–31—that the expression refers to the substratum that undergoes the alteration and not—as I think—to the fact that the qualities that change belong to the same kind of quality.

he already has made it clear that change concerning place and change concerning quantity do not count as alteration. See, for instance, *GC* I.4.319b32–320a2:

> When the change from contrary to contrary is in quantity, it is growth and diminution; when it is in place, it is locomotion; when it is in the affection, i.e., in quality, it is alteration.

So the only cases he has to consider are changes in qualities other than perceptible qualities. In fact, in *Cat.* 8 he distinguishes four kinds of qualities: states and conditions, capacities, affective qualities (perceptible qualities), shapes and forms. As we have seen, in *De generatione et corruptione* and even in *Phys.* V.2.226a26–27, Aristotle does not restrict alteration to change in perceptible qualities. Therefore it is his own former position that is the object of the refutation.

Why does he start with the case of shapes and forms? It seems that this is the most challenging case. Consider the following examples: dough takes on the shape of a star by being molded by a star-shaped cake tin. Bronze takes the shape of a coin by mintage. Bronze takes the shape of Hermes by being in direct contact with the casting mold of Hermes during the casting process. Are these not clear cases of something acquiring a quality by being in direct contact with something that already has this quality? The answer seems to be "yes." However, Aristotle would exclude these cases from his list of true alterations. Why? The only plausible answer seems to be that the shape the substratum loses and the shape it acquires are not contraries lying on a common scale. Thus this change does not fulfill the conditions of alteration explained above.

However, we must recognize that Aristotle does not use this argument in his own refutation. As the passage 245b9–246a9 shows, it rather seems that he assimilates change in shape or form to generation. And generation—as the passage from *De generatione et corruptione* quoted above shows—is not an alteration. This for two reasons: the first—exploited by Ursula Coope in her commentary—is that if a thing comes to be, it cannot be regarded as a permanent substratum of the process it undergoes; in fact, it only exists at the end of the process. If one wants to counter this argument one may say that at least there must be some matter that could be considered as the permanent substratum of the process of generation. For instance, when a bed is put together from its wooden parts, these parts as such persist during the process. They first are potentially a bed and then become actually a bed (see *Metaph.* IX.7). Against this Aristotle could argue that being potentially a bed and being actually a bed are not two contrary qualities that lie on a common scale. Lacking substantial form and having substantial form are not contrary qualities, let alone perceptible qualities.

The case of change in states occupies the longest part of our chapter. What is Aristotle's reason for excluding these cases? As we have seen, in *De generatione et corruptione* he accepted change from health to illness as a case of alteration and he explicitly emphasized that health and illness are contraries lying on a common scale. Therefore this time the reason for excluding this kind of change cannot be the fact that the relevant properties are not contraries. What other argument does our interpretation offer? Well, it is evident that nothing acquires a state like health and illness by being in direct contact with something that already has this state.[45] Unfortunately it is not the case that—as Kafka fancies in a famous short story—the ill person gets healthy because the doctor is laid in his bed. However, here again the arguments Aristotle actually uses in our chapter do not directly rely on this point. He argues with the fact that states like health and illness are relations of underlying qualities of the body, therefore alteration occurs in the latter "in their own right" and becoming healthy is only a change depending on this alteration.

---

[45] The case of contamination is not a counterexample, because unlike in the case of heating contamination is not a direct transmission of a quality, but rather an indirect change through the transmission of microbes.

Aristotle, Physics *VII.3.245b3–248a9: Towards a Fuller Synopsis of the Two Versions*

Oliver Primavesi

It is well known that the first three chapters of Aristotle's *Phys.* VII are transmitted in two different versions, α and β, and that Simplicius knew two versions even of the entire seventh book (Simplicius, *Phys.* 1036, 4–6). For the first chapter and for the beginning of the second, Bekker had printed the text of version β only, but Shute (1882) has provided manuscript evidence for the complete α-version, Ross (1936) has printed version α within his running text, and version β as an appendix after book VIII; a summary of the facts and of different attempts at explaining them has been provided by Manuwald (1971), 7–13; Wardy (1990) offers a synopsis of the two versions in Greek (6–29) and in English translation (40–61). In Wardy's synopsis, however, the correspondence between the two versions appears to be weaker than it actually is, since many sentences or paragraphs of version α have a close equivalent in version β, but at another place of the argument. Therefore, it seemed useful to provide a synopsis of the two versions of chapter 3 which attempts to do what Wardy's does not: to show the correspondence between the two versions even in those cases in which the *sequence* of arguments is different because a sentence or paragraph has been *transposed* in one of the two versions. In order to achieve this aim, in the following synopsis those sentences and paragraphs of either version which do not come at the same stage of the argument as in the other version, have been printed *twice*: they will appear (i) in normal type at the place where they are transmitted, and (ii) in smaller type and in square brackets at the place where the other version offers a corresponding sentence or paragraph.

VERSION α (HIbcjy)	VERSION β (EF[I]JK)
(**245**b3) Ὅτι δὲ τὸ ἀλλοιούμενον ἅπαν ἀλλοιοῦται ὑπὸ τῶν αἰ-(b4)σθητῶν, καὶ ἐν μόνοις ὑπάρχει τούτοις ἀλλοίωσις ὅσα καθ'(b5) αὑτὰ λέγεται πάσχειν ὑπὸ τῶν αἰ-σθητῶν, ἐκ τῶνδε θεωρη-(b6)τέον.	(**245**b19) Ὅτι δὲ τὰ ἀλλοιούμενα ἀλλοιοῦνται πάντα ὑπὸ τῶν αἰσθητῶν, καὶ (b20) μόνων τούτων ἔστιν ἀλλοίωσις ὅσα καθ' αὑτὰ πάσχει ὑπὸ τούτων, (b21) ἐκ τῶνδε θεωρήσωμεν.
τῶν γὰρ ἄλλων μάλιστ' ἄν τις ὑπολάβοι ἔν τε τοῖς σχή-(b7)μασι καὶ ταῖς μορφαῖς καὶ ἐν ταῖς ἕξεσι καὶ ταῖς τούτων (b8) λήψεσι καὶ ἀποβολαῖς ἀλλοίωσιν ὑπάρχειν	τῶν γὰρ ἄλλων μάλιστα [ἄν τις ὑπολάβοι ἔν τε] τοῖς (b22) σχήμασι καὶ ταῖς μορφαῖς καὶ ταῖς ἕξεσι καὶ ταῖς τούτων ἀποβολαῖς (b23) καὶ λήψεσιν [ἀλλοίωσιν ὑπάρχειν.]
ἐν οὐδετέροις δ' (b9) ἔστιν.	δοκεῖ [γὰρ] ὑπάρχειν τὸ τῆς ἀλλοιώσεως, (b24) οὐκ ἔστιν δὲ οὐδ' ἐν τούτοις,
[cf. **246**a6 ἀλλὰ γίγνεσθαι μὲν ἴσως ἕκαστον ἀναγ-(a7)καῖον ἀλλοιουμένου τινός, οἷον τῆς ὕλης πυκνουμένης ἢ μα-(a8)νουμένης ἢ θερμαινομένης ἢ ψυχομένης, οὐ μέντοι τὰ γιγνό-(a9)μενά γε ἀλλοιοῦται, οὐδ' ἡ γένεσις αὐτῶν ἀλλοίωσίς ἐστιν.]	ἀλλὰ γίγνεται [τὸ σχῆμα] ἀλλοιουμένων τινῶν (b25) ταῦτα (πυκνουμένης γὰρ ἢ μανουμένης ἢ θερμαινομένης ἢ ψυχομένης τῆς (b26) ὕλης), ἀλλοίωσις δὲ οὐκ ἔστιν.
τὸ μὲν γὰρ σχηματιζόμενον καὶ ῥυθμιζόμενον ὅταν ἐπι-(b10)τελεσθῇ,	ἐξ οὗ μὲν γὰρ ἡ μορφὴ τοῦ ἀνδριάντος,
οὐ λέγομεν ἐκεῖνο ἐξ οὗ ἐστιν, οἷον τὸν ἀνδριάντα χαλ-(b11)κὸν ἢ τὴν πυραμίδα κηρὸν ἢ τὴν κλίνην ξύλον,	οὐ λέ-(b27)γομεν τὴν μορφήν, οὐδ' ἐξ οὗ τὸ σχῆμα τῆς πυραμίδος ἢ τῆς κλίνης,
ἀλλὰ παρω-(b12)νυμιάζοντες τὸ μὲν χαλκοῦν, τὸ δὲ κήρινον, τὸ δὲ ξύλινον. τὸ (b13) δὲ πεπονθὸς καὶ ἠλλοιωμένον προσαγορεύομεν·	ἀλλὰ (b28) παρωνυμιάζοντες τὸ μὲν χαλκοῦν τὸ δὲ κήρινον τὸ δὲ ξύλινον. τὸ δ' ἀλ-(b29)λοιούμενον λέγομεν·
ὑγρὸν γὰρ (b14) καὶ θερμὸν καὶ σκληρὸν τὸν χαλκὸν λέγομεν καὶ τὸν κηρόν (b15) (καὶ οὐ μόνον οὕτως, ἀλλὰ καὶ τὸ ὑγρὸν καὶ τὸ θερμὸν (b16) χαλκὸν λέγομεν), ὁμωνύμως τῷ πάθει προσαγορεύοντες τὴν (**246**a1) ὕλην.	τὸν γὰρ χαλκὸν ὑγρὸν εἶναι λέγομεν ἢ θερμὸν ἢ σκλη-(**246**a21)ρόν (καὶ οὐ μόνον οὕτως, ἀλλὰ καὶ τὸ ὑγρὸν καὶ τὸ θερμὸν χαλκόν), (a22) ὁμωνύμως λέγοντες τῷ πάθει τὴν ὕλην.
ὥστ' εἰ κατὰ μὲν τὸ σχῆμα καὶ τὴν μορφὴν οὐ λέγεται (a2) τὸ γεγονὸς ἐν ᾧ ἐστι τὸ σχῆμα,	ἐπεὶ οὖν ἐξ οὗ μὲν ἡ μορφὴ καὶ (a23) τὸ σχῆμα καὶ τὸ γεγονὸς ὁμωνύμως οὐ λέγεται τοῖς ἐξ ἐκείνου σχήμασιν,

κατὰ δὲ τὰ πάθη καὶ τὰς (a3) ἀλλοιώσεις λέγεται, φανερὸν ὅτι οὐκ ἂν εἶεν αἱ γενέσεις (a4) ἀλλοιώσεις. ἔτι δὲ καὶ εἰπεῖν οὕτως ἄτοπον ἂν δό-(a5)ξειεν, ἠλλοιῶσθαι τὸν ἄνθρωπον ἢ τὴν οἰκίαν ἢ ἄλλο ὁτιοῦν (a6) τῶν γεγενημένων·	(a24) τὸ δ' ἀλλοιούμενον τοῖς πάθεσιν ὁμωνύμως λέγεται, φανερὸν ὡς ἐν μόνοις (a25) τοῖς αἰσθητοῖς ἡ ἀλλοίωσις. ἔτι καὶ ἄλλως ἄτοπον. τὸ γὰρ λέγει τὸν (a26) ἄνθρωπον ἠλλοιῶσθαι ἢ τὴν οἰκίαν λαβοῦσαν τέλος γελοῖον,
[cf. (**246**a17) ὥσπερ οὖν οὐδὲ τὸ τῆς οἰκίας τελείωμα λέγομεν (a18) ἀλλοίωσιν (ἄτοπον γὰρ εἰ ὁ θριγκὸς καὶ ὁ κέραμος ἀλ-(a19)λοίωσις, ἢ εἰ θριγκουμένη καὶ κεραμουμένη ἀλλοιοῦται ἀλλὰ μὴ τελειοῦται ἡ οἰκία).]	εἰ τὴν τελείω-(a27)σιν τῆς οἰκίας, τὸν θριγκὸν ἢ τὴν κεραμίδα, φήσομεν ἀλλοίωσιν εἶναι, <ἢ> θριγ-(a28)κουμένης τῆς οἰκίας ἢ κεραμιδουμένης ἀλλοιοῦσθαι τὴν οἰκίαν, δῆλον δὴ (a29) ὅτι τὸ τῆς ἀλλοιώσεως οὐκ ἔστιν ἐν τοῖς γιγνομένοις.
ἀλλὰ γίγνεσθαι μὲν ἴσως ἕκαστον ἀναγ-(a7)καῖον ἀλλοιουμένου τινός, οἷον τῆς ὕλης πυκνουμένης ἢ μα-(a8)νουμένης ἢ θερμαινομένης ἢ ψυχομένης, οὐ μέντοι τὰ γιγνό-(a9)μενά γε ἀλλοιοῦται, οὐδ' ἡ γένεσις αὐτῶν ἀλλοίωσίς ἐστιν.	[cf. **245**b24 ἀλλὰ γίγνεται [τὸ σχῆμα] ἀλλοιουμένων τινῶν (b25) ταῦτα (πυκνουμένης γὰρ ἢ μανουμένης ἢ θερμαινομένης ἢ ψυχομένης τῆς (b26) ὕλης), ἀλλοίωσις δὲ οὐκ ἔστιν.]
(a10) ἀλλὰ μὴν οὐδ' αἱ ἕξεις οὔθ' αἱ τοῦ σώματος οὔθ' αἱ τῆς ψυ-(a11)χῆς ἀλλοιώσεις. αἱ μὲν γὰρ ἀρεταὶ αἱ δὲ κακίαι τῶν (a12) ἕξεων· οὐκ ἔστι δὲ οὔτε ἡ ἀρετὴ οὔτε ἡ κακία ἀλλοίωσις,	οὐδὲ γὰρ ἐν ταῖς ἕξε-(a30)σιν. αἱ γὰρ ἕξεις ἀρεταὶ καὶ κακίαι,
(a13) ἀλλ' ἡ μὲν ἀρετὴ τελείωσίς τις (ὅταν γὰρ λάβῃ τὴν αὐτοῦ (a14) ἀρετήν, τότε λέγεται τέλειον ἕκαστον – τότε γὰρ ἔστι μάλιστα (a15) [τὸ] κατὰ φύσιν – ὥσπερ κύκλος τέλειος, ὅταν μάλιστα (a16) γένηται κύκλος καὶ ὅταν βέλτιστος), ἡ δὲ κακία φθορὰ (a17) τούτου καὶ ἔκστασις·	[cf. **246**b27 ἡ μὲν (b28) γὰρ ἀρετὴ τελείωσίς τις (ἕκαστον γὰρ τότε μάλιστα τέλειόν ἐστιν. ὅταν (b29) τύχῃ τῆς οἰκείας ἀρετῆς, καὶ μάλιστα κατὰ φύσιν, καθάπερ ὁ κύκλος (b30) τότε μάλιστα κατὰ φύσιν ἐστίν, ὅταν μάλιστα κύκλος ᾖ), ἡ δὲ κακία (**247**a20) φθορὰ τούτων καὶ ἔκστασις.]
ὥσπερ οὖν οὐδὲ τὸ τῆς οἰκίας τελείωμα λέγομεν (a18) ἀλλοίωσιν (ἄτοπον γὰρ εἰ ὁ θριγκὸς καὶ ὁ κέραμος ἀλ-(a19)λοίωσις, ἢ εἰ θριγκουμένη καὶ κεραμουμένη ἀλλοιοῦται ἀλλὰ (a20) μὴ τελειοῦται ἡ οἰκία),	[cf. **247**a25 τὸ γὰρ λέγει τὸν (a26) ἄνθρωπον ἠλλοιῶσθαι ἢ τὴν οἰκίαν λαβοῦσαν τέλος γελοῖον, εἰ τὴν τελείω-(a27)σιν τῆς οἰκίας, τὸν θριγκὸν ἢ τὴν κεραμίδα, φήσομεν ἀλλοίωσιν εἶναι, <ἢ> θριγ-(a28)κουμένης τῆς οἰκίας ἢ κεραμιδουμένης ἀλλοιοῦσθαι τὴν οἰκίαν, δῆλον δὴ (a29) ὅτι τὸ τῆς ἀλλοιώσεως οὐκ ἔστιν ἐν τοῖς γιγνομένοις.]
τὸν αὐτὸν τρόπον καὶ ἐπὶ τῶν ἀρε-(**246**b1)τῶν καὶ τῶν κακιῶν καὶ τῶν ἐχόντων ἢ λαμβανόντων· αἱ (b2) μὲν γὰρ τελειώσεις αἱ δὲ ἐκστάσεις εἰσίν, ὥστ' οὐκ ἀλλοιώ-(b3)σεις.	

ἔτι δὲ καί φαμεν ἁπάσας εἶναι τὰς ἀρετὰς ἐν τῷ (b4) πρός τι πὼς ἔχειν.

ἀρετὴ δὲ πᾶσα καὶ κακία τῶν (**246**b21) πρός τι,

τὰς μὲν γὰρ τοῦ σώματος, οἷον ὑγίειαν (b5) καὶ εὐεξίαν, ἐν κράσει καὶ συμμετρίᾳ θερμῶν καὶ ψυχρῶν (b6) τίθεμεν,

καθάπερ ἡ μὲν ὑγίεια θερμῶν καὶ ψυχρῶν συμμετρία τις,

ἢ αὐτῶν πρὸς αὐτὰ τῶν ἐντὸς ἢ πρὸς τὸ περιέχον·

ἢ τῶν (b22) ἐντὸς ἢ πρὸς τὸ περιέχον.

(b7) ὁμοίως δὲ καὶ τὸ κάλλος καὶ τὴν ἰσχὺν καὶ τὰς ἄλλας (b8) ἀρετὰς καὶ κακίας. ἑκάστη γάρ ἐστι τῷ πρός τι πὼς ἔχειν,

ὁμοίως δὲ καὶ τὸ κάλλος καὶ ἡ ἰσχὺς τῶν (b23) πρός τι.

(b9) καὶ περὶ τὰ οἰκεῖα πάθη εὖ ἢ κακῶς διατίθησι τὸ ἔχον·

διαθέσεις γάρ τινες τοῦ βελτίστου πρὸς τὸ ἄριστον,

(b10) οἰκεῖα δ' ὑφ' ὧν γίγνεσθαι καὶ φθείρεσθαι πέφυκεν.

λέγω δὲ τὸ (b24) βέλτιστον τὸ σῶζον καὶ διατιθὲν περὶ τὴν φύσιν.

ἐπεὶ οὖν (b11) τὰ πρός τι οὔτε αὐτά ἐστιν ἀλλοιώσεις, οὔτε ἔστιν αὐτῶν ἀλ-(b12)λοίωσις οὐδὲ γένεσις οὐδ' ὅλως μεταβολὴ οὐδεμία, φανερὸν (b13) ὅτι οὔθ' αἱ ἕξεις οὔθ' αἱ τῶν ἕξεων ἀποβολαὶ καὶ λήψεις (b14) ἀλλοιώσεις εἰσίν, ἀλλὰ γίγνεσθαι μὲν ἴσως αὐτὰς καὶ φθείρεσθαι (b15) ἀλλοιουμένων τινῶν ἀνάγκη, καθάπερ καὶ τὸ εἶ-(b16)δος καὶ τὴν μορφήν, οἷον θερμῶν καὶ ψυχρῶν ἢ ξηρῶν καὶ (b17) ὑγρῶν, ἢ ἐν οἷς τυγχάνουσιν οὖσαι πρώτοις.

ἐπεὶ οὖν αἱ μὲν ἀρεταὶ (b25) καὶ αἱ κακίαι τῶν πρός τι, ταῦτα δὲ οὔτε γενέσεις εἰσὶν οὔτε γένεσις αὐ-(b26)τῶν οὐδ' ὅλως ἀλλοίωσις, φανερὸν ὡς οὐκ ἔστιν ὅλως τὸ τῆς ἀλλοιώσεως (b27) περὶ τὰς ἕξεις.

περὶ ταῦτα γὰρ (b18) ἑκάστη λέγεται κακία καὶ ἀρετή, ὑφ' ὧν ἀλλοιοῦσθαι πέ-(b19)φυκε τὸ ἔχον· ἡ μὲν γὰρ ἀρετὴ ποιεῖ ἢ ἀπαθὲς ἢ ὡδὶ (b20) παθητικόν, ἡ δὲ κακία παθητικὸν ἢ ἐναντίως ἀπαθές.

[cf. **247**a22 ἡ μὲν ἀρετὴ ἤτοι ἀπάθειά τις ἢ παθη-(a23)τικὸν ὡδί, ἡ δὲ κακία παθητικὸν ἢ ἐναντία πάθησις τῇ ἀρετῇ.]

ὁμοίως (**247**a1) δὲ καὶ ἐπὶ τῶν τῆς ψυχῆς ἕξεων·

οὐδὲ δὴ περὶ τὰς τῆς ψυχῆς ἀρετὰς καὶ κακίας.

[cf. **246**a13 ἀλλ' ἡ μὲν ἀρετὴ τελείωσίς τις (ὅταν γὰρ λάβῃ τὴν αὑτοῦ (a14) ἀρετήν, τότε λέγεται τέλειον ἕκαστον – τότε γὰρ ἔστι μάλιστα (a15) [τὸ] κατὰ φύσιν – ὥσπερ κύκλος τέλειος, ὅταν μάλιστα (a16) γένηται κύκλος καὶ ὅταν βέλτιστος), ἡ δὲ κακία φθορὰ τούτου καὶ (a17) ἔκστασις.]

ἡ μὲν (b28) γὰρ ἀρετὴ τελείωσίς τις (ἕκαστον γὰρ τότε μάλιστα τέλειόν ἐστιν. ὅταν (b29) τύχῃ τῆς οἰκείας ἀρετῆς, καὶ μάλιστα κατὰ φύσιν, καθάπερ ὁ κύκλος (b30) τότε μάλιστα κατὰ φύσιν ἐστίν, ὅταν μάλιστα κύκλος ᾖ), ἡ δὲ κακία (**247**a20) φθορὰ τούτων καὶ ἔκστασις.

ἅπασαι γὰρ καὶ αὗται (a2) τῷ πρός τι πὼς ἔχειν, καὶ αἱ μὲν ἀρεταὶ τελειώσεις, αἱ (a3) δὲ κακίαι ἐκστάσεις. ἔτι δὲ ἡ μὲν ἀρετὴ εὖ διατίθησι πρὸς τὰ (a4) οἰκεῖα πάθη, ἡ δὲ κακία κακῶς.

ὥστ' οὐδ' αὗται ἔσονται (a5) ἀλλοιώσεις· οὐδὲ δὴ αἱ ἀποβολαὶ καὶ αἱ λήψεις αὐτῶν. (a6) γίγνεσθαι δ' αὐτὰς ἀναγκαῖον ἀλλοιουμένου τοῦ αἰσθητικοῦ μέ-(a7)ρους. ἀλλοιωθήσεται δ' ὑπὸ τῶν αἰσθητῶν· [cf. **246**b19 ἡ μὲν γὰρ ἀρετὴ ποιεῖ ἢ ἀπαθὲς ἢ ὡδὶ (b20) παθητικόν, ἡ δὲ κακία παθητικὸν ἢ ἐναντίως ἀπαθές.]	 γίγνεται μὲν οὖν ἀλλοιουμένου τινὸς ἡ λῆψις (a21) τῆς ἀρετῆς καὶ ἡ τῆς κακίας ἀποβολή, ἀλλοίωσις μέντοι τούτων οὐδέτερον. (a22) ὅτι δ' ἀλλοιοῦταί τι, δῆλον. ἡ μὲν γὰρ ἀρετὴ ἤτοι ἀπάθειά τις ἢ παθη-(a23)τικὸν ὡδί, ἡ δὲ κακία παθητικὸν ἢ ἐναντία πάθησις τῇ ἀρετῇ.
ἅπασα γὰρ ἡ ἠθικὴ (a8) ἀρετὴ περὶ ἡδονὰς καὶ λύπας τὰς σωματικάς,	καὶ τὸ (a24) ὅλον τὴν ἠθικὴν ἀρετὴν ἐν ἡδοναῖς καὶ λύπαις εἶναι συμβέβηκεν·
αὗται δὲ ἢ (a9) ἐν τῷ πράττειν ἢ ἐν τῷ μεμνῆσθαι ἢ ἐν τῷ ἐλπίζειν. αἱ (a10) μὲν οὖν ἐν τῇ πράξει κατὰ τὴν αἴσθησίν εἰσιν, ὥσθ' ὑπ' αἰ-(a11)σθητοῦ τινὸς κινεῖσθαι, αἱ δ' ἐν τῇ μνήμῃ καὶ ἐν τῇ ἐλ-(a12)πίδι ἀπὸ ταύτης εἰσίν· ἢ γὰρ οἷα ἔπαθον μεμνημένοι ἥδονται, (a13) ἢ ἐλπίζοντες οἷα μέλλουσιν. ὥστ' ἀνάγκη πᾶσαν τὴν τοιαύτην (a14) ἡδονὴν ὑπὸ τῶν αἰσθητῶν γίγνεσθαι. ἐπεὶ δ' ἡδονῆς καὶ λύ-(a15)πης ἐγγιγνομένης καὶ ἡ κακία καὶ ἡ ἀρετὴ ἐγγίγνεται (περὶ (a16) ταύτας γάρ εἰσιν), αἱ δ' ἡδοναὶ καὶ αἱ λῦπαι ἀλλοιώσεις (a17) τοῦ αἰσθητικοῦ, φανερὸν ὅτι ἀλλοιουμένου τινὸς ἀνάγκη καὶ (a18) ταύτας ἀποβάλλειν καὶ λαμβάνειν. ὥσθ' ἡ μὲν γένεσις (a19) αὐτῶν μετ' ἀλλοιώσεως, αὐταὶ δ' οὐκ εἰσὶν ἀλλοιώσεις.	ἢ γὰρ (a25) κατ' ἐνέργειαν τὸ τῆς ἡδονῆς ἢ διὰ μνήμην ἢ ἀπὸ τῆς ἐλπίδος. εἰ μὲν οὖν (a26) κατ' ἐνέργειαν, αἴσθησις τὸ αἴτιον, εἰ δὲ διὰ μνήμην ἢ δι' ἐλπίδα, ἀπὸ (a27) ταύτης· ἢ γὰρ οἷα ἐπάθομεν μεμνημένοις τὸ τῆς ἡδονῆς ἢ οἷα πεισόμεθα (a28) ἐλπίζουσιν.
(**247**b1) ἀλλὰ μὴν οὐδ' αἱ τοῦ νοητικοῦ μέρους ἕξεις ἀλλοιώσεις, οὐδ' (b2) ἔστιν αὐτῶν γένεσις. πολὺ γὰρ μάλιστα τὸ ἐπιστῆμον ἐν τῷ (b3) πρός τι πὼς ἔχειν λέγομεν.	ἀλλὰ μὴν οὐδ' <ἐν> τῷ διανοητικῷ μέρει τῆς ψυχῆς ἀλλοίωσις. (a29) τὸ γὰρ ἐπιστῆμον μάλιστα τῶν πρός τι λέγεται.
ἔτι δὲ καὶ φανερὸν ὅτι οὐκ ἔστιν (b4) αὐτῶν γένεσις· τὸ γὰρ κατὰ δύναμιν ἐπιστῆμον οὐδὲν αὐτὸ (b5) κινηθὲν ἀλλὰ τῷ ἄλλο ὑπάρξαι γίγνεται ἐπιστῆμον. ὅταν (b6) γὰρ γένηται τὸ κατὰ μέρος, ἐπίσταταί πως τὰ καθόλου τῷ (b7) ἐν μέρει.	τοῦτο δὲ δῆλον· κατ' οὐδε-(a30)μίαν γὰρ δύναμιν κινηθεῖσιν ἐγγίγνεται τὸ τῆς ἐπιστήμης, ἀλλ' ὑπάρξαντός (**247**b20) τινος· ἐκ γὰρ τῆς κατὰ μέρος ἐμπειρίας τὴν καθόλου λαμβάνομεν ἐπιστή-(b21)μην.
πάλιν δὲ τῆς χρήσεως καὶ τῆς ἐνεργείας οὐκ ἔστι (b8) γένεσις,	οὐδὲ δὴ ἡ ἐνέργεια γένεσις,
εἰ μή τις καὶ τῆς ἀναβλέψεως καὶ τῆς ἀφῆς (b9) οἴεται γένεσιν εἶναι· τὸ γὰρ χρῆσθαι καὶ τὸ ἐνεργεῖν ὅμοιον τούτοις.	εἰ μή τις καὶ τὴν ἀνάβλεψιν καὶ τὴν (b22) ἀφὴν γενέσεις φησίν· τοιοῦτον γὰρ ἡ ἐνέργεια.

ἡ δ' ἐξ ἀρ-(b10)χῆς λῆψις τῆς ἐπιστήμης γένεσις οὐκ ἔστιν οὐδ' ἀλλοίωσις· τῷ γὰρ ἠρεμῆ-(b11)σαι καὶ στῆναι τὴν διάνοιαν ἐπίστασθαι καὶ φρονεῖν λεγόμεθα,
(b12) εἰς δὲ τὸ ἠρεμεῖν οὐκ ἔστι γένεσις· ὅλως γὰρ οὐδεμιᾶς μετα-(b13)βολῆς, καθάπερ εἴρηται πρότερον.
ἔτι δ' ὥσπερ ὅταν ἐκ τοῦ (b14) μεθύειν ἢ καθεύδειν ἢ νοσεῖν εἰς τἀναντία μεταστῇ τις, οὔ (b15) φαμεν ἐπιστήμονα γεγονέναι πάλιν (καίτοι ἀδύνατος ἦν τῇ (b16) ἐπιστήμῃ χρῆσθαι πρότερον),
οὕτως οὐδ' ὅταν ἐξ ἀρχῆς λαμ-(b17)βάνῃ τὴν ἕξιν·
τῷ γὰρ καθίστασθαι τὴν ψυχὴν ἐκ τῆς φυ-(b18) σικῆς ταραχῆς φρόνιμόν τι γίγνεται καὶ ἐπιστῆμον.
διὸ καὶ (b19) τὰ παιδία οὔτε μανθάνειν δύνανται
οὔτε κατὰ τὰς αἰσθήσεις (**248**a1) ὁμοίως κρίνειν τοῖς πρεσβυτέροις·
πολλὴ γὰρ ἡ ταραχὴ (a2) καὶ ἡ κίνησις.
καθίσταται δὲ καὶ ἠρεμίζεται
πρὸς ἔνια μὲν (a3) ὑπὸ τῆς φύσεως αὐτῆς,
πρὸς ἔνια δ' ὑπ' ἄλλων,
ἐν ἀμ-(a4)φοτέροις δὲ ἀλλοιουμένων τινῶν τῶν ἐν τῷ σώματι,
καθά-(a5)περ ἐπὶ τῆς χρήσεως καὶ τῆς ἐνεργείας, ὅταν νήφων γένη-(a6)ται καὶ ἐγερθῇ.
φανερὸν οὖν ἐκ τῶν εἰρημένων
ὅτι τὸ ἀλλοι-(a7)οῦσθαι καὶ ἡ ἀλλοίωσις
ἔν τε τοῖς αἰσθητοῖς γίγνεται
καὶ ἐν (a8) τῷ αἰσθητικῷ μορίῳ τῆς ψυχῆς,
ἐν ἄλλῳ δ' οὐδενὶ πλὴν (a9) κατὰ συμβεβηκός.

ἡ δὲ ἐξ ἀρχῆς λῆψις τῆς (b23) ἐπιστήμης οὐκ ἔστι γένεσις οὐδ' ἀλλοίωσις· τῷ γὰρ ἠρεμίζεσθαι καὶ καθ-(b24)ίστασθαι τὴν ψυχὴν ἐπιστήμων γίγνεται καὶ φρόνιμος.
καθάπερ οὖν οὐδ' ὅταν (b25) καθεύδων ἐγερθῇ τις ἢ μεθύων παύσηται ἢ νοσῶν καταστῇ, γέγονεν ἐπι-(b26)στήμων·
καίτοι πρότερον οὐκ ἐδύνατο χρῆσθαι καὶ κατὰ τὴν ἐπιστήμην ἐνερ-(b27)γεῖν,
εἶτα ἀπαλλαγείσης τῆς ταραχῆς καὶ εἰς ἠρεμίαν καὶ κατάστασιν ἐλθούσης τῆς (b28) διανοίας ὑπῆρξεν ἡ δύναμις ἡ πρὸς τὴν τῆς ἐπιστήμης χρείαν.
τοιοῦτο δή (b29) τι γίγνεται καὶ τὸ ἐξ ἀρχῆς ἐν τῇ τῆς ἐπιστήμης ὑπαρχῇ·
τῆς γὰρ ταρα-(b30)χῆς ἠρεμία τις καὶ κατάστασις.
οὐδὲ δὴ τὰ παιδία δύναται μαθεῖν
οὐδὲ (**248**a26) κρίνειν ταῖς αἰσθήσεσιν ὁμοίως τοῖς πρεσβυτέροις.
πολλὴ γὰρ ἡ ταραχὴ (a27) περὶ ταῦτα καὶ ἡ κίνησις.
καθίσταται δὲ καὶ παύεται τῆς ταραχῆς
τοτὲ (a28) μὲν ὑπὸ τῆς φύσεως
τοτὲ δ' ὑπ' ἄλλων.
ἐν ἀμφοτέροις δὲ τούτοις ἀλ-(**248**b26)λοιοῦσθαί τι συμβαίνει,
καθάπερ ὅταν ἐγερθῇ καὶ γένηται νήφων πρὸς τὴν (b27) ἐνέργειαν.
φανερόν οὖν
ὅτι τὸ τῆς ἀλλοιώσεως
ἐν τοῖς αἰσθητοῖς
καὶ ἐν τῷ (b28) αἰσθητικῷ μέρει τῆς ψυχῆς,
ἐν ἄλλῳ δ' οὐθενὶ πλὴν κατὰ συμβεβηκός.

Bibliography

Ackrill, J. L., *Aristotle's Categories and De interpretatione* (Oxford: Clarendon Press, 1963).

Barnes, J. and Bonelli, M., eds., *Aristote: physique et métaphysique* (Paris: forthcoming).

Barnes, J., ed., *Revised Oxford Translation* (Princeton: Princeton University Press, 1984).

Bekker, I., *Aristotelis graece, ex recognitione Immanuelis Bekkeri* (Berlin: Academia Regia Borussica, 1831–1870).

Bodéüs, R., *Aristote, Catégories* (Paris: Les Belles Lettres, 2001).

Boeri, M. D., *Aristoteles. Fisica, libros VII–VIII,* traducción, introducción y commentario (Buenos Aires: Biblos, 2003).

Carteron, H., *Aristote. Physique,* texte établi et traduit par Henri Carteron, 2 vol. (Paris: Les Belles Lettres, 1926–1931).

Cornford, F. M., and Wicksteed, P. H., *The Physics, with an English Translation,* 2. vol. (Cambridge, MA: Harvard University Press, 1934).

Couloubaritsis, L., *L'avènement de la science physique. Essai sur la Physique d'Aristote* (Brussels: Ousia, 1980).

Davidson, D., *Essays on Actions and Events* (Oxford: Clarendon Press, 1980. Reprint Oxford: Oxford University Press, 2001).

De Haas, F. A. J., "The Discriminating Capacity of the Soul in Aristotle's Theory of Learning," in Ricardo Salles, ed., *Metaphysics, Soul and Ethics in Ancient Thought. Themes from the Work of Richard Sorabji* (Oxford: Oxford University Press, 2005), 321–344.

Diels, H., ed., *Simplicii in Aristotelis Physicorum libros quattuor posteriores commentaria*, CAG X (Berlin: G. Reimer, 1895. Reprint Berlin: De Gruyter, 1962).

Frede, M., *Essays in Ancient Philosophy* (Oxford: Clarendon Press, 1987).

Frede, M. and Patzig, G., *Aristoteles Metaphysik Z: Text, Übersetzung und Kommentar* (München: C. H. Beck, 1988).

Gosling, J. C. B. and Taylor, C. C. W., *The Greeks on Pleasure* (Oxford: Clarendon Press, 1982).

Hagen, Ch., *Simplicius, On Aristotle Physics 7*, translated by Charles Hagen (London: Duckworth and Ithaca: Cornell University Press, 1994).

Hardie, R. P. and Gaye, R. K., *Translation of the* Physics in William D. Ross, ed., *The Complete Works of Aristotle*, vol. 2 (Oxford: Clarendon Press, 1930).

Judson, L., *Aristotle's Physics: A Collection of Essays* (Oxford: Oxford University Press, 1995).

Kalbfleisch, C., ed., *Simplicii in Aristotelis Categorias commentarium*, CAG VIII (Berlin: G. Reimer, 1907).

Kim, J., *Supervenience and Mind: Selected Philosophical Essays* (Cambridge: Cambridge University Press, 1993).

Lettinck, P. and Urmson, J. O., *Philoponus, On Aristotle's* Physics *5–8 with Simplicius, on Aristotle on the Void*, translated by Paul Lettinck and J. O. Urmson (Ithaca: Cornell University Press 1994).

Lloyd, A. C., "The Principle that the Cause is Greater than its Effect," *Phronesis* 21 (1976), 146–156.

Magee, J. M., "Sense Organs and the Activity of Sensation in Aristotle," *Phronesis* 45 (2000), 121–149.

Mansion, A., *Introduction à la physique aristotélicienne* (Louvain-Paris: J. Vrin, 1945; 2nd edition, revised and augmented, 1946).

Manuwald, B., *Das Buch* H *der aristotelischen* Physik. *Eine Untersuchung zur Einheit und Echtheit* (Meisenheim am Glan: A. Hain, 1971).

Mourelatos, A. P. D., "Quality, Structure, and Emergence in Later Pre-Socratic Philosophy," in J. Cleary, ed., *Proceeedings of the Boston Area Colloquium in Ancient Philosophy*, vol. II (Ithaca: University Press of America, 1987), 127–194.

Nagel, E., *The Structure of Science, Problems in the Logic of Scientific Explanation* (Burlingame: Harcourt, Brace and World, 1961).

Owen, G. E. L., *Logic, Science, and Dialectic. Collected Papers in Greek Philosophy*, edited by M. Nussbaum (London: Duckworth and Ithaca: Cornell University Press, 1986).

Pellegrin, P., *Aristote. Physique,* traduction et présentation par Pierre Pellegrin (Paris: Flammarion, 2000).

Politis, V., *Aristotle and the Metaphysics* (London and New York: Routledge, 2004; e-Library, London: Taylor & Francis Group, 2005).

Prantl, C., *Aristoteles' Acht Bücher Physik* (Leipzig: Engelmann, 1854).

Ross, W. D., *Aristotle's Physics.* A revised text with introduction and commentary by W. D. Ross (Oxford: Clarendon Press, 1936).

Ross, W. D., *Aristotelis Physica*, recognovit brevique adnotatione critica instruxit W. D. Ross, OCT (Oxford: Clarendon Press, 1950; Reprint with corrections, Oxonii: E Typographeo Clarendoniano, 1973).

Ruggiu, L., *Aristotele: Fisica*, saggio introduttivo, trad., note e apparati di Luigi Ruggiu, 2 ed., (Milano: Mimesis, Rusconi, 2007).

Russo, A., *Aristotele: Fisica, Del cielo*, trad. di A. R. e O. Longo, 8 ed., (Roma-Bari: Laterza, 2005).

Sedley, D., "Aristotelian Relativities," in M. Canto-Sperber and P. Pellegrin, eds., *Le style de la pensée. Recueil de textes en hommage à Jacques Brunschwig* (Paris: Les Belles Lettres, 2002), 324–352.

Seel, G., *Die Aristotelische Modaltheorie* (Berlin: Walter de Gruyter, 1982).

Shute, R., *Aristotle's* Physics *Book VII. A Transcript of the Paris Ms. 1859 collated with the Paris Mss. 1861 and 2683 and a Manuscript in the Bodleian Library with an Introductory Account of these Manuscripts* (Oxford: Clarendon Press, 1882).

Sorabji, R. R. K., "Aristotle and the Instant of Change," in J. Barnes, M. Schofield and R. Sorabji, eds., *Articles on Aristotle 3:* Metaphysics (London: Duckworth, 1979), 159–177.

Tracy, Th. J., *Physiological Theory and the Doctrine of the Mean in Plato and Aristotle* (The Hague: Mouton, 1969).

Verbeke, G., "L'argument du livre VII de la *Physique*. Une impasse philosophique," 1969; reprinted in: Id., *D'Aristote à Thomas d'Aquin* (Leuven: Leuven University Press, 1990), 250–267.

Vitelli, H., *Iohannis Philoponi in Aristotelis libros quinque posteriores comm.*, ed. H. Vitelli, CAG XVII (Berlin: 1888).

Wagner, H., *Aristoteles. Physikvorlesung*, übersetzt von Hans Wagner, 4 Aufl. (Berlin: Akademic Verlag, 1983).

Wardy, R., *The Chain of Change: A Study of Aristotle's* Physics *VII* (Cambridge: Cambridge University Press, 1990).

Waterlow, S., *Nature, Change and Agency in Aristotle's Physics: A Philosophical Study* (Oxford: Oxford University Press, 1981).

Wieland, W., *Die Aristotelische Physik* (Göttingen: Vandenhoek & Rupprecht, 1962).

Woods, M., "Universals and Particular Forms in Aristotle's *Metaphysics*," *Oxford Studies in Ancient Philosophy*, suppl. vol.: *Aristotle and the Later Tradition* (1991), 41–56.

Index Locorum

General Index

Index of Greek Terms

• ALSO AVAILABLE FROM PARMENIDES PUBLISHING •

PRE-SOCRATICS

By Being, It Is: *The Thesis of Parmenides* by Néstor-Luis Cordero
To Think Like God: *Pythagoras and Parmenides. The Origins of Philosophy.* Scholarly and fully annotated edition by Arnold Hermann
The Illustrated To Think Like God: *Pythagoras and Parmenides. The Origins of Philosophy by* Arnold Hermann with over 200 full color illustrations.
The Legacy of Parmenides: *Eleatic Monism and Later Presocratic Thought* by Patricia Curd
Parmenides and the History of Dialectic: *Three Essays* by Scott Austin
Parmenides, Venerable and Awesome: *Proceedings of the International Symposium* edited by Néstor-Luis Cordero
Plato's Parmenides: *Text, Translation & Introductory Essay by Arnold Hermann. Translation in collaboration with Sylvana Chrysakopoulou and a Foreword by Douglas Hedley*
The Route of Parmenides: *Revised and Expanded Edition, With a New Introduction, Three Supplemental Essays, and an Essay by Gregory Vlastos* by Alexander P. D. Mourelatos
The Fragments of *Parmenides: A Critical Text with Introduction and Translation, the Ancient* Testimonia *and a Commentary* by A. H. Coxon. Revised and Expanded Edition edited with new Translations by Richard McKirahan and a new Preface by Malcolm Schofield

PLATO

God and Forms in Plato by Richard D. Mohr
Image and Paradigm in Plato's Sophist by David Ambuel
Interpreting Plato's Dialogues by J. Angelo Corlett
One Book, the Whole Universe: *Plato's Timaeus Today* edited by Richard D. Mohr and Barbara M. Sattler
The Philosopher in Plato's Statesman by Mitchell Miller
Platonic Patterns: *A Collection of Studies* by Holger Thesleff
Plato's Late Ontology: *A Riddle Resolved* by Kenneth M. Sayre
Plato's Parmenides: *Text, Translation & Introductory Essay by Arnold Hermann. Translation in collaboration with Sylvana Chrysakopoulou and a Foreword by Douglas Hedley*
Plato's Universe by Gregory Vlastos

ARISTOTLE

One and Many in Aristotle's *Metaphysics—Volume I: Books Alpha-Delta* by Edward C. Halper

One and Many in Aristotle's *Metaphysics—Volume 2: The Central Books* by Edward C. Halper

Reading Aristotle:* Physics *VII.3 "What is Alteration?" *Proceedings of the International ESAP-HYELE Conference* edited by Stefano Maso, Carlo Natali, and Gerhard Seel

HELLENISTIC PHILOSOPHY

A Life Worthy of the Gods*: The Materialist Psychology of Epicurus* by David Konstan

ETHICS

Sentience and Sensibility*: A Conversation about Moral Philosophy* by Matthew R. Silliman

AUDIOBOOKS

The Iliad (unabridged) by Stanley Lombardo
The Odyssey (unabridged) by Stanley Lombardo
The Essential Homer by Stanley Lombardo
The Essential Iliad by Stanley Lombardo

FORTHCOMING

Plotinus,* Ennead *IV.8: On the Soul's Descent into Body
Translation with an Introduction & Commentary by Barrie Fleet

***A Stranger's Knowledge: Statesmanship, Philosophy, and Law in Plato's* Statesman** by Xavier Marquez

Presocratics and Plato*: A Festschrift in Honor of Charles Kahn* edited by Arnold Hermann, Vassilis Karasmanis and Richard Patterson

ANNOUNCING

A NEW SERIES OF COMMENTARIES ON
THE ENNEADS OF PLOTINUS:
Translations with Introductions and Philosophical Commentaries on Individual Treatises
Series edited by John M. Dillon and Andrew Smith